Environmental Chemistry and Recent Pollution Control Approaches

Edited by Hugo Saldarriaga-Noreña, Mario Alfonso Murillo-Tovar, Robina Farooq, Rajendra Dongre and Sara Riaz

Published in London, United Kingdom

IntechOpen

Supporting open minds since 2005

Environmental Chemistry and Recent Pollution Control Approaches
http://dx.doi.org/10.5772/intechopen.80247
Edited by Hugo Saldarriaga-Noreña, Mario Alfonso Murillo-Tovar, Robina Farooq, Rajendra Dongre and Sara Riaz

Contributors
Moufti Ahmed, Katsutoshi Inoue, Hari Paudyal, Keisuke Ohto, Hidetaka Kawakita, Liming Shen, Chengyun Feng, Yan Wei, Sijian Xia, Hua Zhang, Danqing Zhao, Fang Yao, Xukun Liu, Yuxi Zhao, Huajie Zhang, Dionisio Rodríguez, Jan Broda, Francesca Bosco, Chiara Mollea, Magdalena Urbaniak, Elżbieta Mierzejewska, Cristina Navarro, Antonio Leyva, Cristian Mateo, Micaela Navarro, Medhat Rehan, Abdullah S. Alsohim, Shuisen Chen, Ming Zhong, Luqman Ali Shah, Sher Ali Khan

First published in London, United Kingdom, 2019 by IntechOpen
IntechOpen is the global imprint of INTECHOPEN LIMITED, registered in England and Wales, registration number: 11086078, 7th floor, 10 Lower Thames Street, London, EC3R 6AF, United Kingdom
Printed in Croatia

British Library Cataloguing-in-Publication Data
A catalogue record for this book is available from the British Library

Additional hard and PDF copies can be obtained from orders@intechopen.com

Environmental Chemistry and Recent Pollution Control Approaches
Edited by Hugo Saldarriaga-Noreña, Mario Alfonso Murillo-Tovar, Robina Farooq, Rajendra Dongre and Sara Riaz
p. cm.
Print ISBN 978-1-83968-062-5
Online ISBN 978-1-83968-063-2
eBook (PDF) ISBN 978-1-83968-064-9

For EU product safety concerns:
IN TECH d.o.o., Prolaz Marije Krucifikse Kozulić 3, 51000 Rijeka, Croatia, info@intechopen.com or visit our website at intechopen.com.

Meet the editors

Dr. Hugo Saldarriaga-Noreña currently works at the Autonomous University of Morelos State Cuernavaca, México. He received his Bachelor in Chemistry in 1991 from the University of Antioquia (Medellín-Colombia), his Master of Science in Environmental Chemistry (2003) and Ph.D. (2008) from the Autonomous University of Mexico (UNAM). Dr. Saldarriaga-Noreña's research areas are: preparation of samples of biological and/or environmental origin; assessment of agro-industrial waste; degradation of organic compounds through the application of advanced oxidation processes; analysis of organic compounds (pesticides, aliphatic hydrocarbons, polycyclic aromatic hydrocarbons, and their nitroderivatives) and metals using gas chromatography coupled to mass spectrometry, liquid chromatography, and atomic absorption, respectively; validation of instrumental analysis techniques; chemometrics, statistical analysis of data using univariate and multivariate techniques.

Dr. Mario Alfonso Murillo-Tovar is a Research Environmental Chemist at the Universidad Autónoma del Estado de Morelos, México. His research has focused on development and validation of analytical methods, chemical characterization of environmental samples, and pollution remediation. His research has been supported by CONACYT, INECC-SEMARNAT, and ISSSTE in México. He is the author and co-author of more than 25 research publications, one book, and three book chapters. At UAEM, he teaches graduate-level and undergraduate courses in environmental chemistry, atmospheric chemistry, quantitative chemical analysis, chemometrics, probability and statistical development of analytical methods, and sample preparation techniques in analytical chemistry. He received his undergraduate Bachelor of Chemistry in 2004 from Universidad del Valle, Colombia. He obtained his Master's degree (2007) and Doctorate in Chemical Sciences (2012) from the Universidad Nacional Autónoma de México.

Dr. Robina Farooq has been involved in teaching, research, management, and academic work in numerous distinguished universities of Britain, China, and Pakistan for the last 28 years. Currently, she is working at COMSATS Institute of Information Technology, Lahore, Pakistan. She has discovered innovative and low-cost processes for the treatment of wastewater. She is the author of scientific manuscripts, books, book chapters, and granted patents by USPTO, USA. She is the recipient of Best Innovator, Best University Teacher, and Productive Scientist Awards. She has worked on projects including ultrasonic decomposition of pollutants, phytoremediation of wastewater, bioelectrochemical synthesis of renewable fuel, bioelectrochemical decomposition of wastewater and energy recovery, recovery of heavy metals from effluents, microbial fuel cell technology for wastewater remediation, and retrieval of precious metals from printed circuit boards.

Rajendra S. Dongre received his M.Sc. from the Department of Chemistry, R.T.M., Nagpur University in 1996 (Gold Medalist) and his PhD in 2010. His research work includes organic synthesis, chitosan bio-composites, matrixes, and remediation of water pollution de-fluoridation, nitrate, chromium, and phosphate lead (II). He has worked as a Scientists-B in CSIR-LAB, National Environmental Engineering Research Institute (NEERI) Nagpur M.S., India. Overall, he has 25 years of experience in research and development and 17 years of post-graduate teaching experience, which has resulted in fifty-five international/national research publications. He has also guided four PhD research students. He received the 6th National Award (runner-up position) for Technology Innovation in Petrochemicals and Downstream Plastics Processing Industry, for research in the field of polymer science and technology, handed by Hon'ble Ananth Kumar, Petrochemical & Fertilizers Minister of Government of India in 2016. He received the 5th National Science & Technology Award for research contribution in the field of developing science in 2017, by EET-CRS, Noida, India.

Dr Sara Riaz is currently working at COMSATS University Islamabad, Lahore campus as an Assistant Professor and is involved in teaching, research, and management in the Department of Chemistry. She completed her MSc from Bahauddin Zakariya University, Multan Pakistan in 2013 in organic synthesis and received a doctoral degree in 2014 from East China University of Science and Technology, Shanghai, P.R. China for her work on synthesis and characterization of novel ubiquinone derivatives and their electrochemical interaction studies with gold nanoparticles. Dr Sara is currently involved in projects including bioelectrochemical decomposition of dyes for electricity generation, synthesis of graphitic nanomaterials and their applications in energy storage devices. and heavy metal ions detection.

Contents

Preface XI

Section 1
Environmental Pollution 1

Chapter 1 3
Characterization of the Youssoufia-Morocco-MineFluoride-Contaminated Water and Their Detrimental Effects on Human Health
by Moufti Ahmed

Chapter 2 21
Greenhouse Gas Emissions of Agriculture: A Comparative Analysis
by Dionisio Rodríguez

Chapter 3 41
Progressive Research in the Molecular Mechanisms of Chronic Fluorosis
by Liming Shen, Chengyun Feng, Sijian Xia, Yan Wei, Hua Zhang, Danqing Zhao, Fang Yao, Xukun Liu, Yuxi Zhao and Huajie Zhang

Section 2
Pollution Control Approaches 67

Chapter 4 69
Polymer Hydrogels for Wastewater Treatment
by Luqman Ali Shah and Sher Ali Khan

Chapter 5 81
Biological Remediation of Phenoxy Herbicide-Contaminated Environments
by Magdalena Urbaniak and Elżbieta Mierzejewska

Chapter 6 103
Biodegradation of Sheep Wool Geotextiles Designed for Erosion Control
by Jan Broda

Chapter 7 123
Adsorptive Removal of Fluoride onto Different Waste Materials: Orange Juice Residue, Waste Seaweed, and Spent Cation-Exchange Resin
by Katsutoshi Inoue, Hari Paudyal, Hidetaka Kawakita and Keisuke Ohto

Chapter 8 145
Bioremediation of Heavy Metals
by Medhat Rehan and Abdullah S. Alsohim

Chapter 9 161
Bioremediation of Petroleum-Contaminated Soil
by Shuisen Chen and Ming Zhong

Chapter 10 173
Mycoremediation in Soil
by Francesca Bosco and Chiara Mollea

Chapter 11 189
Arsenic Phytoremediation: Finally a Feasible Approach in the Near Future
by Cristian Mateo, Micaela Navarro, Cristina Navarro and Antonio Leyva

Preface

Environmental pollution is not a new phenomenon. Different studies are currently being carried out around the world to explain many of the impacts caused by exposure to certain pollutants. However, such studies in the first instance are not carried out at the required speed, which allows the avoidance of some of these impacts. And in other cases, there is no follow-up to the problems caused by pollution, due to the lack of infrastructure and economic resources. The purpose of evaluating the level of contamination in the environment is to find a strategy for mitigating or preventing impacts and/or looking for mechanisms to remediate contaminated matrices. In this sense, this book aims to contribute by presenting new studies aimed at determining levels of environmental pollution in various parts of the world. It also shows new alternatives for the remediation of contaminated matrices.

The editors wish to thank PRODEP (Program for the Development of Teachers) and CONACyT (National Council of Science and Technology) for constant support in the financing of publications.

Hugo Saldarriaga-Noreña and Mario Alfonso Murillo-Tovar
Autonomous University of State of Morelos,
Mexico

Robina Farooq and Sara Riaz
COMSATS University Islamabad,
Pakistan

Rajendra Dongre
RTM Nagpur University,
India

Section 1

Environmental Pollution

Chapter 1

Characterization of the Youssoufia-Morocco-MineFluoride-Contaminated Water and Their Detrimental Effects on Human Health

Moufti Ahmed

Abstract

In Youssoufia, the second phosphate mining center of our country (Morocco), the drinking water needs of the rural population are of underground origins. Indeed, most of Youssoufia's rural areas feed on traditional wells. The main purpose of this chapter is to evaluate the degree of contamination of mine water along the pumping canal by fluoride. Wells located near this channel were also analyzed to see the influence of the existence of black phosphate in this region on these wells. At the end of this analytical part, it is obvious to conclude that the dewatering waters of the black phosphate mines of Youssoufia, known as dewatering water along the canal, contain significant fluoride concentrations in the order of 3–4 mg/l on average and the waters of the wells located near this canal have fluoride concentrations higher than the standard recommended by the National Office of Drinking Water in Morocco and the World Health Organization which is 1.5 mg/l. Indeed, a number of residents residing in Youssoufia suffer from fluorosis.

Keywords: fluoride, characterization, Morocco, mine water, human health

1. Introduction

In Morocco, the establishment of a water resources management policy both quantitatively and qualitatively is not a whim of organization but a necessity for survival. In fact, the reserves of drinking water are not infinite, and many regions suffer either from the lack or the intolerable waste of this vital density.

In Youssoufia, the second phosphate mining center of our country, the drinking water needs of the rural population are of underground origins. Indeed, most rural areas of Youssoufia feed from traditional wells, the closest to each Douar. The drinking water supply is made from four holes of the BAHIRA water table located 24 km southeast of Youssoufia.

On the other hand, there is black phosphate mine water that is not waste water, in the common sense of the term, because it has never actually been used. These are the waters of a multitude of superimposed layers drowning the different layers of phosphates in the southern zone of the Gantour deposits, which requires predigestion to allow the exploitation of these deposits. Currently, this predigestion is carried out essentially in the recipes 7 and 9 where the average flow reaches in 6–8000 m^3/day, and at present, it exceeds 35,000 m^3/day [1].

The main purpose of this chapter is to evaluate the degree of contamination of mine water along the pumping canal by fluoride. Wells located near this channel were also analyzed to see the influence of the existence of black phosphate in this region on these wells.

2. Choice of the zone

In Morocco, the phosphate areas have a geographical area extending from Khouribga, Oued Zem and Tadla to the northeast, Settat to the center, and Ben Guerir and Youssoufia to the west [2]. The choice of the Youssoufia zone is justified on the one hand by the fact that the latter contains significant concentrations of fluoride of the order of 3 mg/l on average in mine water [3–5], and on the other hand, that all the studies that have already been done on regions other than Youssoufia [6, 7].

3. Overview of the study area (Youssoufia)

3.1 Historical aspect

The city of Youssoufia was born in 1930 following the discovery of a phosphate deposit in the Gantour plateau by the French geologist Lois Gentil. Since then, the city took the name of the latter until 1960, when it was renamed Youssoufia by his majesty Mohamed V following the visit he made to this locality on the same date [8].

The development, extension and exploitation of the deposit, as well as the increasing demand for labor, have given the Youssoufia center considerable economic growth and accelerated urbanization.

3.2 Geographic location

The city of Youssoufia is located east of the city of Safi, at a distance of 80 km, north of the city of Marrakech, at a distance of 100 km and south of the city of Casablanca, at a distance of 220 km. It is part of the Doukkala-Abda region, Safi province (**Figure 1**). It is also part of the Gantour plateau, one of the four main deposits of the cherifian office of phosphates (OCP) group, limited to the north by the RAHMANA, to the east by the MOISSAT hills, to the south by the Bahira, and to the west by the TASSAOUT Wadi.

The city of Youssoufia is crossed by a railway that connects the phosphate deposits of Ben Guerir and Youssoufia with the port of SAFI. It is also served by: the Jamaat Shaim road to the northwest, the road to Chemaia on one side and Ben Guerir on the other, the road that joins the main road 12 to Marrakech to the south, and the new road to Sidi Bennour to the north.

Despite this important network of roads and railway lines, the city of Youssoufia remains a locality and on the fringe of the existing clean economic activities on the roads (e.g., Ben Guerir, Sidi Bennour,...) and this is because of the hand of OCP on the city as a single manager concerned with the development-only activities related to the mining sector.

3.3 The climate

The climate of the city of Youssoufia is semicontinental, arid in winter, the average rainfall is about 272 mm per year, and the average minimum temperatures vary between 13°C in January and 25.1°C in July. Highs above 30°C can be observed in the

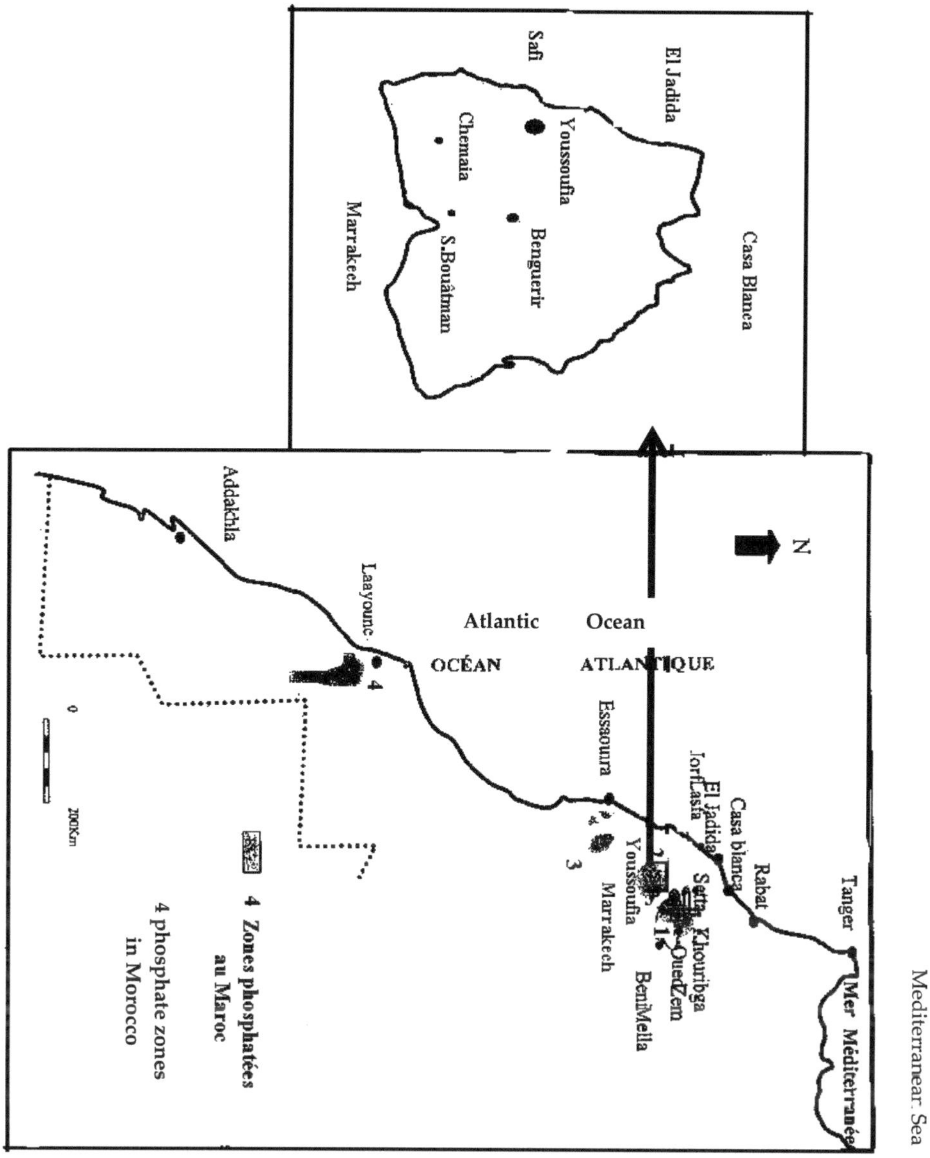

Figure 1.
Geographic location of Youssoufia.

region. The coldest months are December, January, and February when the minimum average is 8.6°C. The winds are relatively weak with predominance of the northern directions during the dry season and the northeast during the wet season [8, 9].

The monthly averages (**Table 1**) show that there is a rainfall accident in November, December, January, and February, which allows a medium or wet year for a crop life until March or April due to the water potential accumulated in the ground. Outside these periods, the existence of a vegetative life is linked to the supply of water by irrigation. The months of July and August are the driest [10, 11].

3.4 Mining potentials

In terms of phosphate, the region is full of a large amount of phosphate; it is the largest reserve of the country after Khouribga. Its reserves are estimated at 2 billion m^3, which are located in the Gantour plateau [12].

Month	Jan.	Feb.	March	April	May	June	July	August	Sept.	Oct.	Nov.	Dec.	Ann.
Average monthly (mm)	39	34	39	22	17	2	0	0	10	25	40	44	272

Table 1.
Average monthly rainfall (Youssoufia station, period 1933–2000).

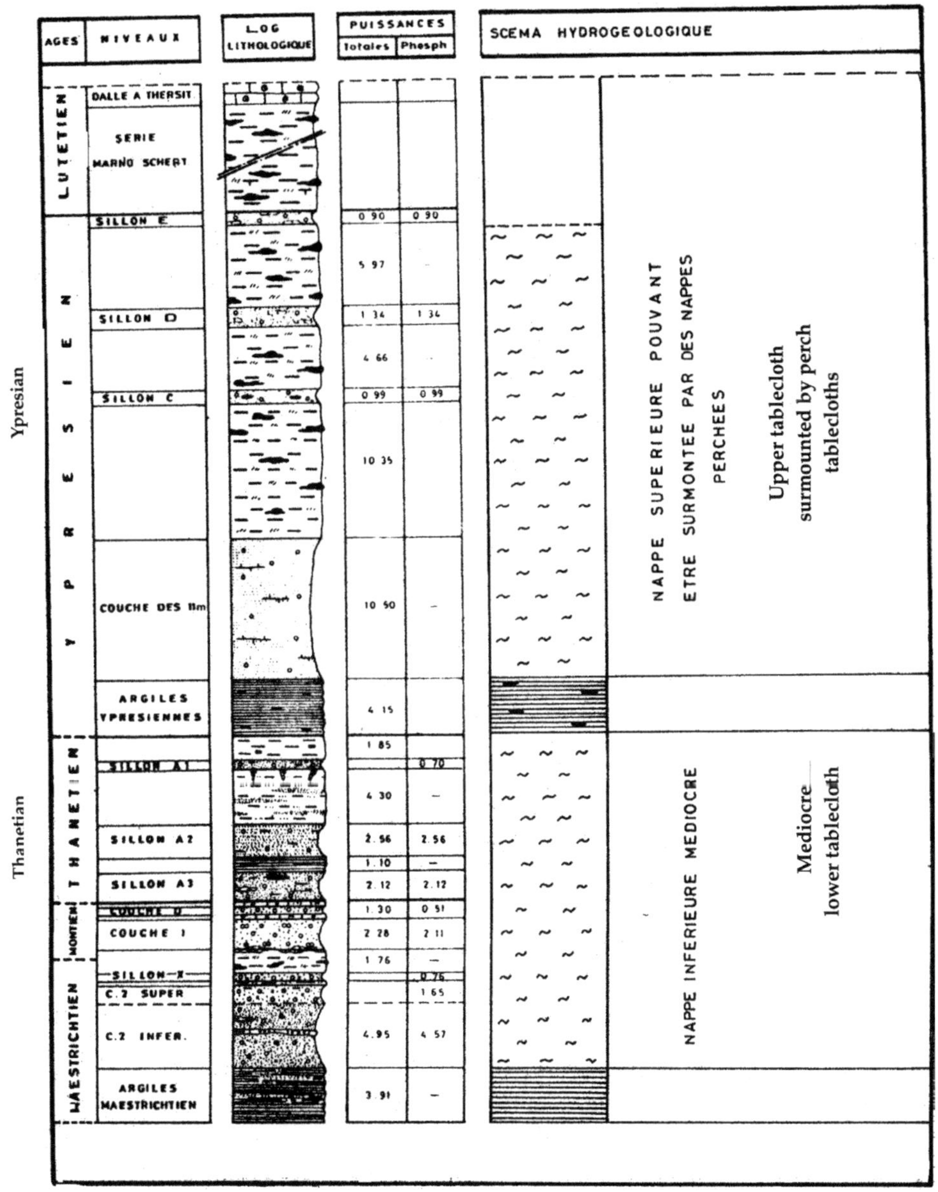

Figure 2.
Average lithological section of the Youssoufia black phosphate deposit.

Phosphate mining is carried out in two areas, clear phosphate and black phosphate. There is a drying unit with a capacity of 60 million tons per year and a black phosphate calcination unit. The O.C.P for some years directs its exploitations on

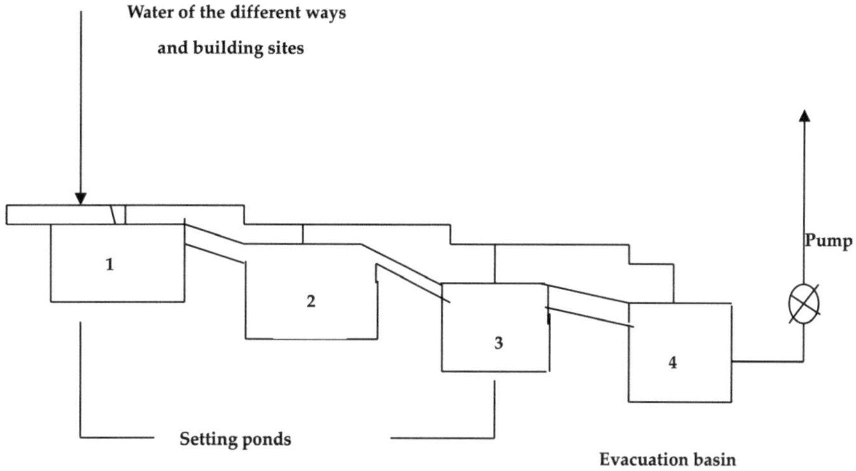

Figure 3.
Mine stations "Exhaure" and design of shallow basins.

Ben Guerir where they are in open sky, when with the city of Youssoufia, they are executed in great depth. This has practically caused a stagnation of the depth since 1970, and a very significant regression since 1984.

3.5 Overview of the hydrogeology of the Youssoufia flood zone

The sedimentary basin is in the form of a subsidence basin, it is a tectonized and folded primary substratum during the New Caledonian tectonic phases and a cover ranging from quaternary to quaternary affected by attenuated tectonic replicas of the tectonic Alpine phase. Currently in the zone of Youssoufia, it is the layer 1 that is in exploitation. It also exploits the layer 0 when the interlayer layer 1-layer 0 is absent or friable. As it moves south, light phosphate is exploited in recipes 5 and 6 (discovery), where it exploits in addition to layer 1 and layer 0, groove X (see **Figure 2**; [10]). The lithological section of this deposit (**Figure 3**) shows that the black phosphate of Youssoufia is located at an average depth of more than 150 m, up to 180 m. The phosphate layer is located below the aquifer, which results in particularly delicate mining problems, which had to be controlled. The exploitation of this deposit is done underground [11].

Layer 1 (Montien): it is a loose sandy phosphate whose power is of the order of 2–2.5 m. The base of this layer is usually in the form of a phosphate limestone. This layer is also surmounted by a cardial spacer. This spacer separates layer 1 from layer 0.

The layer 0 (Montien): it is a loose sandy phosphate less rich in view than the layer 1. The X groove is encountered in the Maastrichtian, which is characterized by an alternation of phosphate levels and sterile levels with predominantly marly levels. The phosphate levels are in the form of phosphated sands called from bottom to top: layer 6, 5, 4, 3, 2, and the groove X. The layers 0 and 1 are in the Montien, Sillon X in the Maastrichtian.

4. Dewatering water (Exhaure water)

Underground mining at Youssoufia began in 1930 with the extraction of the clear, dry phosphate that runs along the northern limit of the deposit, whose work has bypassed the area drowned in its northern and western parts.

Element	P_2O_5	CaO	CO_2	SiO_2	Al_2O_3	F^-
Percentage	32.5	52	3.7	3.9	0.36	3.9

Table 2.
Average composition of black phosphate.

In fact, this zone is characterized on one hand by the existence of organic matter giving the phosphates their blackish color and acting on its content, which necessitates their calcination before delivery [11], and on the other hand by the presence of water in extraction sites from groundwater that have been discovered since 1960 by the exploitation of recipe 7 and then by recipes 8 and 9. This water called "mine water" is pumped to the outside of the mine to allow the exploitation of these deposits [13]. The observation in recent years of the tonnage dewatering discharge relationship predicts a flow of more than 35,000 m^3/day from the year 2000. The average composition of black phosphate is shown in **Table 2** [11].

Extraction of black phosphate in the submerged area of Youssoufia occurs between two underground aquifers. In this zone, there are two types of devices for precleaning the upper water table: surface precleaning (caving and tracing) and spot precleaning (wells and predigestion soundings) [10].

4.1 Exhaure "mine water"

The water that impedes the smooth progress of phosphate extraction in the underground workings of the flooded Youssoufia zone comes from:

- lightning strikes;
- tracing;
- operating wells; and
- precleaning drains.

To ensure the smooth running of extraction, a dewatering infrastructure was set up for the evacuation of these waters to the day. The scheme adopted for evacuation consists of three phases [10]:

First phase: it concerns the collection of water from the various roads and sites by essentially pneumatic pumps installed in the cuvettes. The water will be evacuated to the nearest settling stations (to the headway).

Second phase: the water collected during the first phase is collected in the secondary basins. These waters are discharged in a 50/60 pipe where 80/90 is connected directly to the pipe 6", which is different in the main dewatering station. Typically, these pumps supply intermediate basins, and the main station consists of a succession of settling ponds (**Figures 3** and **4**).

Third phase: this is the main phase of dewatering; all the water from the sector collected in the settling ponds is repressed by electric pumps (GFP, Rotos, Guinard, and Deplechin). This water is discharged to the day through wells in 6 "pipes to be conveyed by gravity in 10" pipes, which reject them away from the farms to avoid any recycling of water. The position of the dewatering stations that force the inflow of water toward the day depends on the hydrogeology of the zone and the dynamics of exploitation [10] (**Figure 4**).

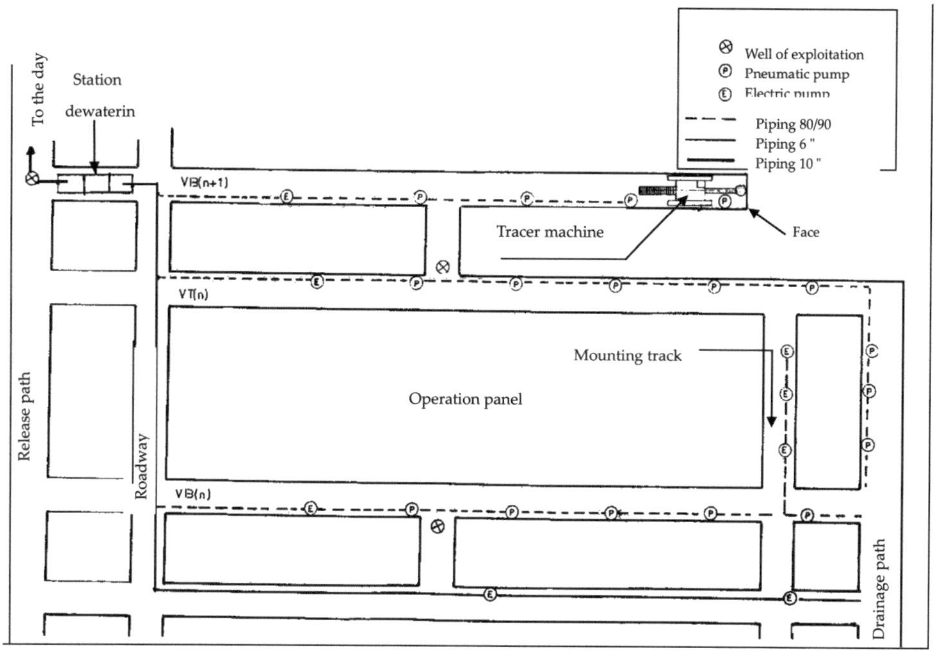

Figure 4.
Operating method.

5. Exhaure and quality

5.1 Characterization techniques for Youssoufia mine water

All the water samples are kept at 4°C in plastic bottles (Year 2001/2002) and analyzed in the laboratory "Water and Environment," at the Faculty of Sciences of El Jadida within 24 h according to the standard French methods (AFNOR) [14, 15]. At each sampling, the temperature, TDS, conductivity, and pH were measured in situ in the field using an HACH conductivity meter, modeled 44,600 and a WTW pH meter, modeled 522 with a combined electrode. Turbidity is measured using a turbidimeter.

In this study, fluoride was assayed by potentiometric method using TISAB solution, a fluoride ion-specific electrode and a reference electrode (Ag/AgCl). It has the advantage of being simple, of rapid response and of being amenable to serial dosages [14, 16]. The apparatus used is a JENCO model 6209 type pH-ionometer equipped with a fluoride-specific electrode.

5.1.1 Nature and methods of taking samples

Samples were taken along the mine-water channel from recipe 9 "unit 9" (Douar OULED ABADE) from stations S1 to S5 and recipes (unit)7 and 8 (station S6) and the mixture of the three recipes located at neighborhood of SIDI AHMED and extend to El BIAR (stations S7, S8, and S9). Samples from wells near the dewatering water channel were also taken (**Figure 5**).

5.1.2 Results of characterization of mine water

Youssoufia mine water from black phosphates corresponds to natural groundwater characterized by a more or less variable chemical composition. **Table 3**

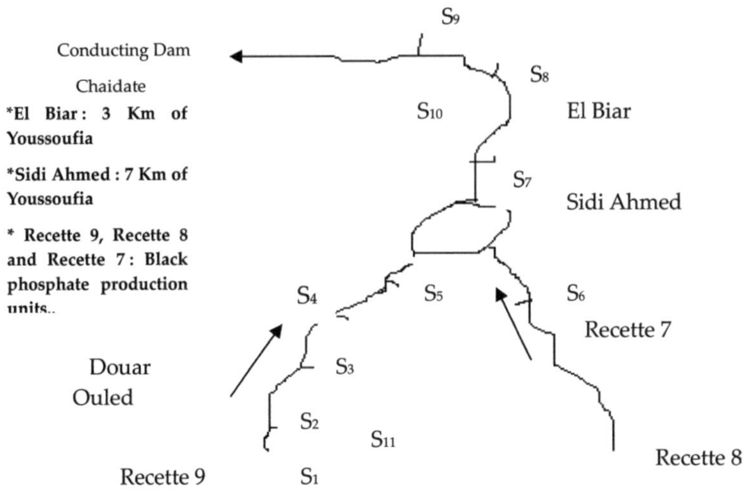

Figure 5.
Dewatering water sampling stations. S10 et S11 : Welles.

summarizes the results of physicochemical analyzes carried out along the canal. It is noted that these waters have electrical conductivities ranging from 1.2 to 1.4 ms/cm at 25°C and pH from 7.67 to 8.22, so we can conclude that these groundwater (mine water) are slightly alkaline.

5.1.2.1 Evolution of the fluoride content for the different stations

Table 4 presents fluoride standards for drinking water as a function of average annual temperature obtained from daytime daily maximums [17].

According to **Table 4**, the world Health Organization (WHO) standard for water with an average temperature of 21.66–27.77°C (average temperature of mine water) is set at 0.8 mg/l. Examination of the results obtained shows that the Youssoufia mine water along the canal has high levels of F^- of the order of 3.5 mg/l on average, thus much higher than the norms published by the WHO. It can be added that these important levels decrease along the canal, moving away from the recipe 9 (station 1) to the mixture of the three recipes 7, 8, and 9 (station 9) located near the village of Charige, the content of F^- decreases from 4.06 to 2.36 mg/l (**Figure 6**). The same results were found by Falgata [1].

This abnormal fluoride content is due to the fact that leaching of the black phosphate from this region contains from 3 to 4% of fluoride [18]. This was well verified in the study conducted by ARAFAN [19] on the monitoring of the release of fluoride ions by black phosphates, which shows that they release between 3 and 3.5 mg/l of fluoride for a stirring time greater than 100 up to 300 h for the different masses used (**Figure 7**). This quantity of fluorides released by black phosphate is greater than that released by white phosphate [20–22].

5.1.2.2 Orthophosphates and magnesium

From station 1 (S1) to station 9 (S9), the contents of orthophosphates (**Figure 8**) and those of magnesium (**Figure 9**) decrease along the dewatering channel, except that in the case of (Mg^{2+}), there is a slight increase in the station 6 (S6) and this is because the mine water of recipes 7 and 8 (station 6) are more concentrated in Mg^{2+}.

Moreover, we can note that the contents of (PO_4^{3-}) and (Mg^{2+}) do not exceed the standards published by WHO [23].

DOI: http://dx.doi.org/10.5772/intechopen.80547

Stations/parameter	S1	S2	S3	S4	S5	S6	S7	S8	S9	S10	S11
pH	7.67	7.73	7.87	7.93	7.99	7.86	7.83	8.10	8.20	7.89	7.92
T (°C)	25.4	25 3	25.3	25.5	25.4	25	25	24.9	24.9	25.3	25.1
E.C (ms/cm)	1.21	1.23	1.25	1.31	1.32	1.34	1.25	1.28	1.25	1.26	1.36
TDS (g/l)	0.58	0.58	0.61	063	0.58	0.58	0.57	0.68	0.65	0.44	0.75
Turbidity	120	120	122	122	122	0	14	0	1	46	0
Cl^- (mg/l)	110.9	120.01	150.1	150.2	200	180	200	220	230	150.1	170
Hardness (mg/l)	135.70	174.85	160.3	158.80	160.4	157.11	154.96	154.96	156.08	140.90	162
Ca^{2+} (mg/l)	48.49	80.19	94.39	98.02	100.7	84.17	94.18	96	100.1	72.14	84.2
Mg^{2+} (mg/l)	87.21	82.66	65.64	60.78	55.37	72.94	60.78	58.07	55.92	68.76	77.8
Nitrate (mg/l)	28	8	14	16	8	28	20	6	4	18	6
Nitrite (mg/l)	0.23	0.011	0.15	0.105	0.064	0	0.01	0	0	0.044	0
AT	0	0	0	0	0	0	0	0	0	0	0
CAT (méq/l)	4.4	4.6	4.8	4.6	4.8	4.3	4.6	5.2	5.4	5.4	6.2
PO_4^{3-} (mg/l)	0.5	0.44	0.38	0.31	0.22	0.2	0.21	0.13	0.03	0.12	0
Oxydabilité (mg/l)	2.8	2.2	1.4	2.2	2	1.4	2.2	2	1.6	1.4	1.6
Fluoride (mg/l)	4.06	3.82	3.4	3.35	3.26	2.76	3.04	2.89	2.36	1.51	1.63

Table 3.
The average of physicochemical parameters of Youssoufia mine water (year 2001/2002).

Average annual temperature (°C) of water from daytime maximums daily	Optimal levels of fluorides (ppm)	Maximum allowable levels of fluorides (ppm)	Limit levels of fluorides (ppm)
10	1	1.074	1.640
10.55	0.956	1.024	1.548
11.11	0.916	0.979	1.465
11.67	0.880	0.940	1.393
12.22	0.849	0.905	1.329
12.78	0.821	0.873	1.270
13.33	0.796	0.844	1.218
13.89	0.773	0.819	1.170
14.45	0.752	0.795	1.127
15	0.733	0.774	1.088
15.55	0.714	0.752	1.048
16.11	0.698	0.734	1.015
16.67	0.682	0.716	0.983
17.22	0.667	0.700	0.953
17.78	0.654	0.685	0.925
18.33	0.640	0.670	0.897
18.89	0.629	0.657	0.874
19.45	0.618	0.644	0.850
20	0.6	0.632	0.28
20.55	0.597	0.621	0.807
21.11	0.587	0.610	/
21.66 to 27.77			0.800
28.33 to 32.22			0.700

Table 4.
The fluoride level standards of drinking water at different temperatures.

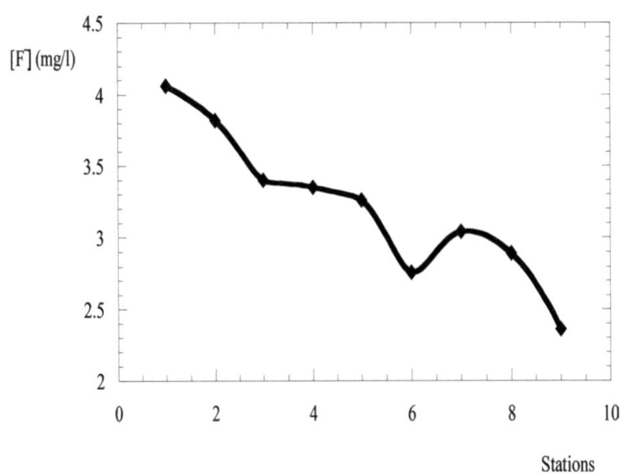

Figure 6.
Evolution of fluorides concentration along the mine water channel.

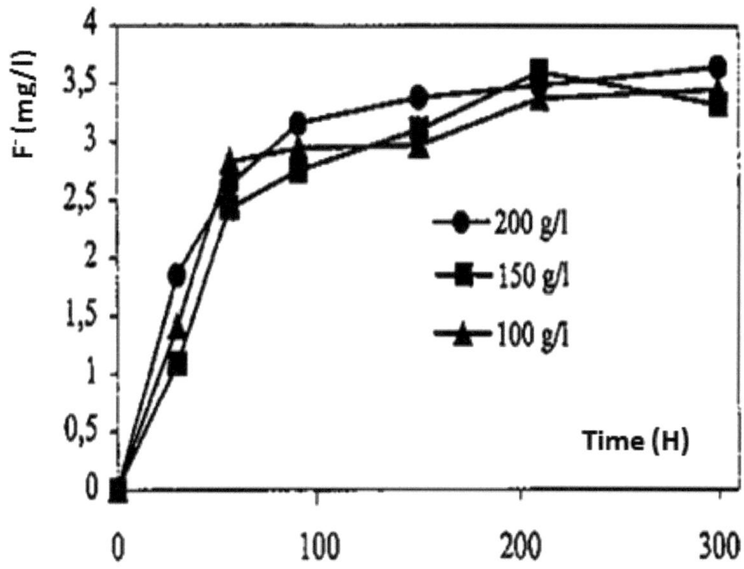

Figure 7.
Releasing fluorides as a function of time to different masses of black phosphate.

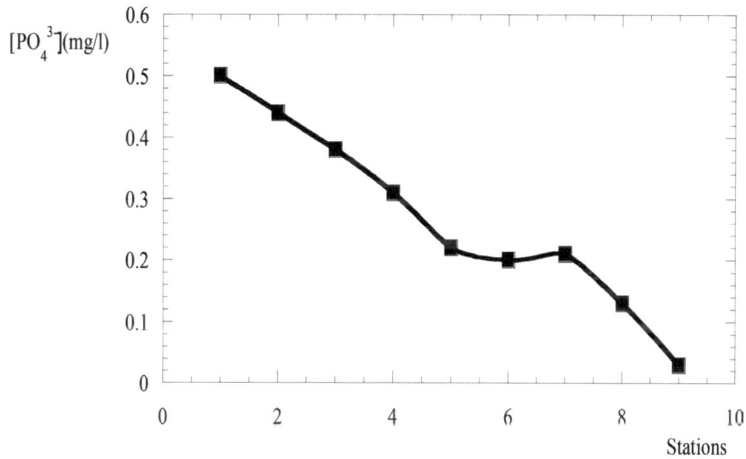

Figure 8.
Evolution of orthophosphates along the canal.

5.1.2.2.1 Relationship between the evolution of concentration of fluorides, magnesium, and orthophosphates concentration

A correlation is made between fluoride (F^-) and orthophosphate (PO_4^{3-}) contents (**Figure 10**) and between fluoride (F^-) and magnesium (Mg^{2+}) (**Figure 11**). The results showed that these three elements decrease in the same way since the correlations is linear. So, they undergo the same phenomenon along the channel, and this can be explained by the precipitation of $\mathbf{Mg_{5-x}\,Cax\,(PO_4)_3\,F, OH}$, which is insoluble in water.

5.1.2.3 Calcium

For calcium, the acceptable limit set by WHO is 75 mg/l. In the study area, mine waters have significant calcium contents, that continue to increase away from station 9 (S9), from station 2 (S2), toward the mixture of the three recipes R7, R8 and

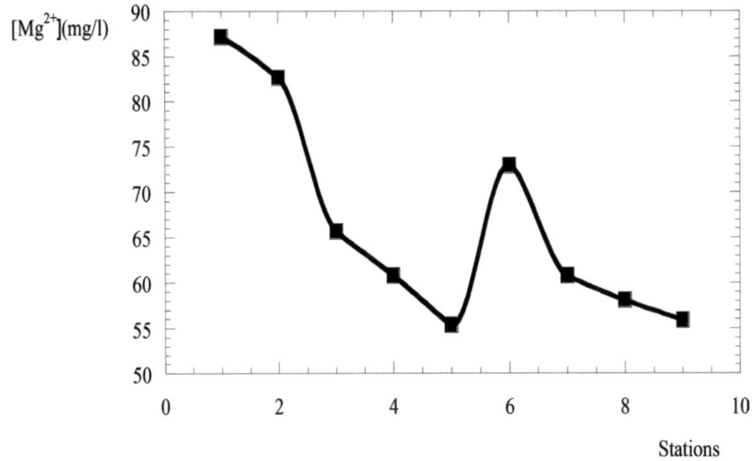

Figure 9.
Evolution of magnesium along the canal.

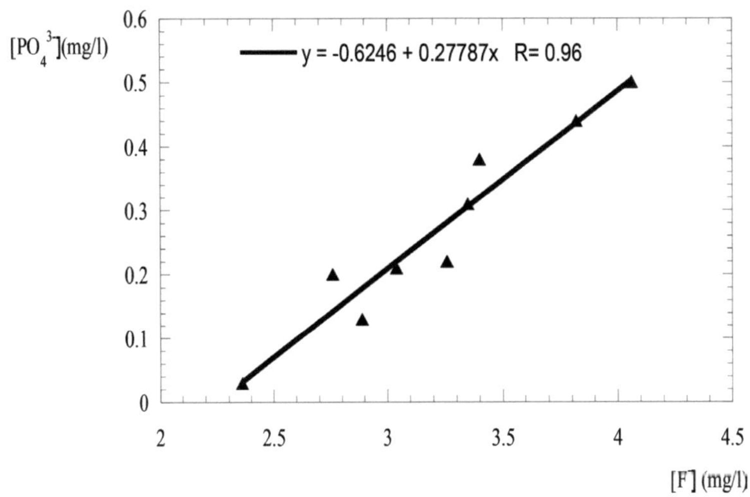

Figure 10.
Relationship between the evolution of concentration of fluorides and orthophosphates.

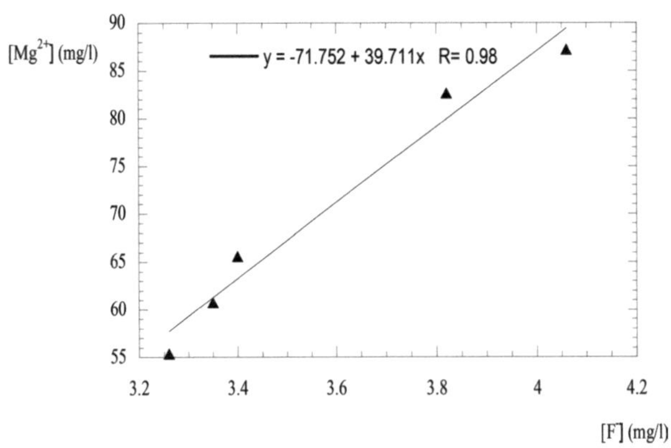

Figure 11.
Relationship between the evolution of concentration of fluoride and magnesium.

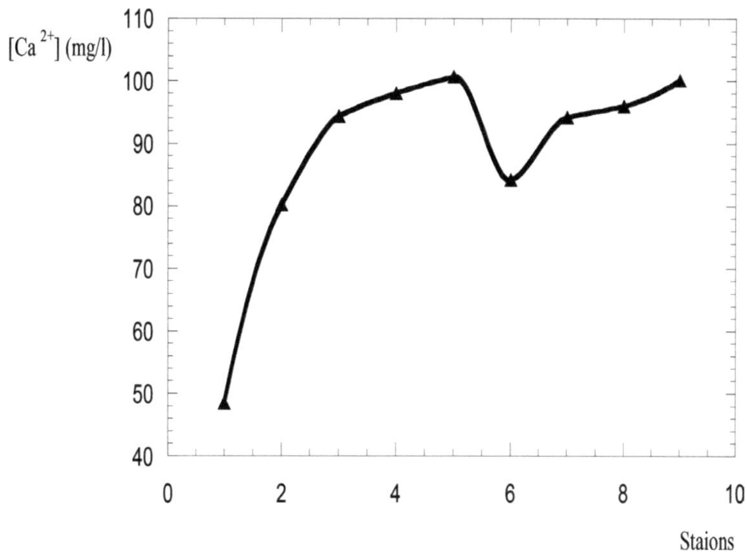

Figure 12.
Evolution of calcium levels along the canal.

R9 (**Figure 12**). These levels are higher than the standard published by WHO. The low value of Calcium in station 6 (S6) is due to the fact that the dewatering water of recipe 7 (R7) is not very well concentrated in this element.

5.1.3 Impact of the phosphatic layer on the groundwater

In order to study the impact of the phosphatic layers on the water table, we sampled two wells (stations 10 and 11) located near the dewatering channel.

To better interpret the results, we have drawn figures that represent the concentration of the element considered in stations 5, 6, and 7 and we compared them with that of stations 10 and 11 to better make a comparison between dewatering water of recipe 9 before mixing (station 5), water of recette 7 and 8 (station 6), the mixture of the three recipes "recette" (station 7), and the well waters (stations 10 and 11). The results are as follows.

The results of the physicochemical analyzes carried out on the water of the well in question showed that these waters are contaminated by several elements, which influences their conductivity (CE) and the hardness (TH) (**Figures 13** and **14**). We can therefore conclude that the phosphatic layer influences the wells (S10 and S11), since the black phosphatic minewater has concentrations with a much higher hardness ($135 < [TH] < 175$ mg/l), this is due may be leaching phosphatic soils. The same remark is noted for conductivity.

5.1.3.1 Fluoride concentration

Figure 15 shows that the well waters (S10 and S11) are contaminated with fluoride; the concentration of fluoride concentration [F^- (S_{10}) = 1.51 and F^- (S_{11}) = 1.63] is greater than the limit set by the WHO of 1.5 mg/l, which is in good agreement with previous results obtained by Mr. Mountadar and his collaborators [20] in a study of Khouribga well waters have shown that these waters are contaminated with fluoride [(F^-) > 1.5 mg/l], and also by Hassani et al. [19], who showed that the well waters in the vicinity of the Youssoufia calcination units have fluoride

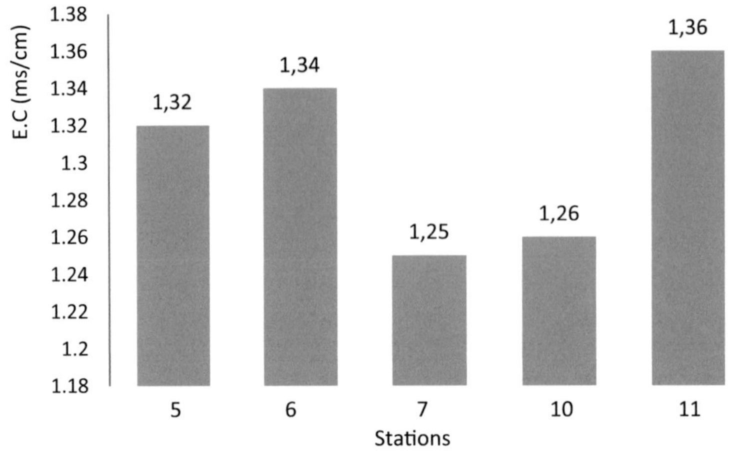

Figure 13.
Comparison between the conductivity of the waters along the canal and that of neighboring wells.

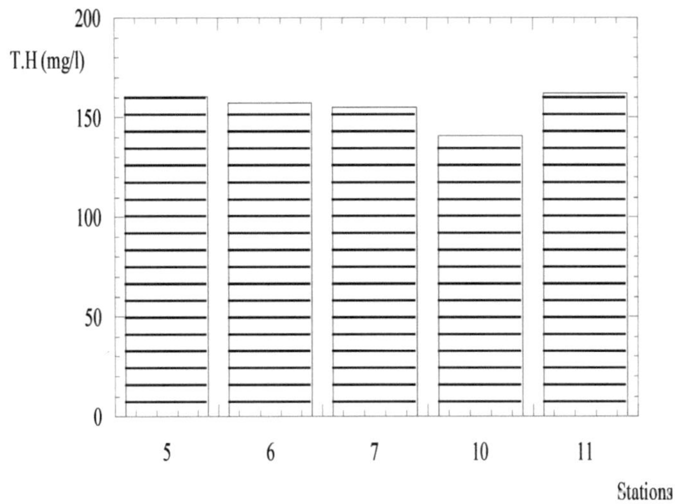

Figure 14.
Evolution of hardness in the study area.

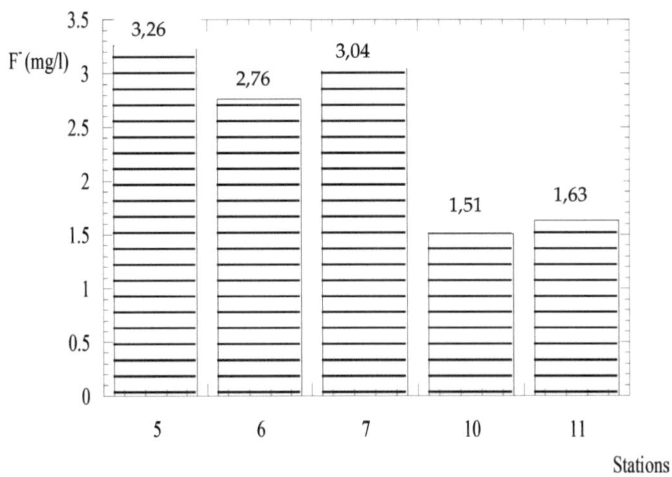

Figure 15.
Comparison between fluoride levels in mine water and neighboring wells.

concentrations between 1.5 and 2.8 mg/l. The consumption of such fluoride-laden water has negative health effects: dental fluorosis and skeletal disorders [3].

For the other elements, it can be said that they are comparable to those obtained for Youssoufia mine water and they do not exceed the standards set by the World Health Organization (WHO).

6. Conclusion

At the end of this analytical part, it is obvious to conclude that:

- Youssoufia mine water contains high concentrations of fluorides of the order of 3–4 mg/l on average; this is due to the fact that the black phosphate of Youssoufia contains 3–4% of F^-. This abnormal fluoride content is due to the fact that leaching of the black phosphate from this region contains 3–4% of F^- [3]. This was well verified in the study conducted by Garmes [2] on the monitoring of the release of fluoride ions by black phosphates, which shows that they release between 3 and 3.5 mg/l of for a stirring time greater than 100 h up to 300 h for the different masses used. This quantity of fluorides released by black phosphate is greater than that released by white phosphate [20].

- The ions (F^-), (Mg^{2+}), and (PO_4^{3--}) are eliminated in the same way along the dewatering channel. This is probably due to the precipitation of magnesium orthophosphates.

- The waters of the wells near the canal are contaminated by fluoride ($F^- \rightarrow 1.5$ mg/l), and the same result was observed by Hassani and his collaborators [24] in a study made on wells located in the vicinity of the Youssoufia calcination units, which showed that these wells have F^- concentrations of between 1.5 and 2.8 mg/l. Indeed, a number of residents reside in Youssoufia suffer from fluorosis. Therefore, there is a permanent risk for the population using the groundwater near the phosphate mining plants in Youssoufia. In addition, Fandi [3] noted during a study (questionnaire) carried out on the inhabitants of the city of Youssoufia that:

    1. Caries is the most common dental disease and is more prevalent in urban than in rural areas, meaning that the diet consumed in cities can also affect the condition of the teeth.

    2. With regard to the state of the teeth, people from rural areas seem to be the most affected; in fact 36% of these subjects have a form of hyperplastic tooth compared to 9.67% of urban people, 45.33 and 49.33% of rural people, respectively, have yellow and rusty pigmentation against 40.32 and 38.71%, respectively, for urban people.

    3. For bone disorders, they are present in 61.3% in urban areas while this figure reaches 77.33% in rural areas.

We, therefore, note the main health problem resulting from the possibility of using mine water as drinking water where irrigation resides in the clinical manifestation of mottled enamel called “Darmous” in Morocco and “Disease of the factories.” In France [3, 8], this causes psychological problems for thousands of people residing in the phosphate zones.

Acknowledgements

Thank you to all my teachers and colleagues from the Water and Environment Laboratory, Department of Chemistry, Faculty of Sciences, Chouaib Doukhali University, El Jadida, Morocco.

Author details

Moufti Ahmed[1,2]

1 Regional Center for the Professions of Education and Training, Settat, Morocco

2 Laboratory for Water and the Environment, Faculty of Sciences, Department of Chemistry, El Jadida, Morocco

*Address all correspondence to: amoufti@gmail.com

References

[1] Falgata S. Les possibilités de traitement et l'utilisation des Eaux d'Exhaure du phosphate noir de Youssoufia. Revue Marocaine du Génie Civil. 1990;**27**(Janvier):5-9

[2] Garmes H. L'analyse des interférences naturelles et anthropiques sur le développement de la fluorose (province de Khouribga); Défluoruration par des procédés conventionnelles et membranaires [Thèse de Doctorat d'état]. El Jadida, Maroc: Université Chouaib Doukkali, Faculté des Sciences; 2002

[3] Fandi R. Etude de la relation entre les teneurs en fluor au niveau de l'eau, du sol et des urines humaines dans les régions Darmous (Youssoufia) [Thèse de Doctorat en pharmacie]. Université Mohammed V, Faculté de Médecine et de Pharmacie de Rabat; 1994

[4] Mountadar M, Garmes H, Moufti A. Défluoruration d'une eau souterraineYoussoufia-Maroc'. Annales Chimie Science Matériaux. 2001;**26**:S341-S344

[5] Kau PMH, Smith DW, Binning P. Fluoride retention by kaolin clay. Contaminant Hydrology. 1997;**28**:89-108

[6] Moufti A, Mountadar M. Valorisation d'une cendre volante par la défluoruration des eaux souterraines, cas de Youssoufia-Maroc. In: 2éme colloque de Groupe Marocain de Recherche pour l'Environnement (GMRE) et le XXXII éme Congrès du Groupe Français des Pesticides (GFP); Mai, Marrakech, Maroc. 2002. pp. 29-31

[7] Groupe Office Chérifien des Phosphates. Rapport de présentation d'activité du groupe OCP. Réalisation Technique-SDG/MG/I; 1987

[8] Yousrani K. Enquête épidémiologique sur l'évolution de la fluorose dentaire de la fluorose dentaire dans la province de Khouribga [Thèse médecine]. Casablanca, Maroc: Université Hassan II; 1990

[9] Mountadar M, Garmes H, Bouraji M. Défluoruration d'une eau chargée en fluorures : Cas du rejet de la laverie des phosphates (Khouribga-Maroc). In: Acte du colloque International Membranes et Procédés de Séparation (CIMP). Vol. 1. 1999. pp. 89-94

[10] Royaume du Maroc, Ministre des Travaux Publics et des Communications, Direction de l'hydraulique, Division des Ressources en Eau. Ressources en eau du Maroc. In: Plaines et bassins du Maroc atlantique, édition du service géologique du Maroc, Rabat. 1975;**2**:367-392

[11] Groupe Office Chérifien des Phosphates. Bultein d'information, no. 54. Décembre; 1992. pp. 7-13

[12] Groupe Office Chérifien des Phosphates, Direction de la Formation et de la Communication, Division Formation et Perfectionnement du Personnel OE P TAMCA: Technique de dénoyage; 1993

[13] Oumhijane S. Recherche documentaire: Géologie et hydrogéologie de la zone noyée de Youssoufia (recensement et synthèse). Rapport de fin de stage effectué à Office Chérifienne du Phosphate (O.C.P) Youssoufia; 1999

[14] Association Française De Normalisation. Qualité de l'eau. Tome 1: Terminologie, échantillonnage et évaluation des méthodes. 3e édition. Paris, France; 1997

[15] DEGRÉMONT SA. Mémento technique de l'eau, édition du cinquantenaire, 9e édition. LAVOISIER (Éditeur), Paris, France, tomes 1 et 2; 1989

[16] Rodier J, Bazin C, Broutin JP, Chambon P, Champsaur H, et Rodi L. L'analyse de l'eau, 8e édition. Dunod (Éditeur), Paris, France; 1996

[17] AFNOR. Eau, Méthodes d'essais. 1er édition. Paris, France; 1979

[18] O.M.S. Fluor et santé. In: Série de monographie. Genève: Organisation mondiale de la Santé; 1972. p. 59

[19] Arafan A, Erraji M, Hassani E, Chik A. Traitement Thermique d'un Phosphate Très Riche En Matières Organiques Et Valorisation du Phosphate Calcine Y2. Rapport nom édité du Groupe OCP, Maroc, Maroc-Phosphore et Cerphos, 30 Juin; 1998

[20] Mountadar M, Garmes H, Bouraji M, Lhadi EK. Défluoruration d'une eau chargée en fluorures: cas du rejet de la laverie des phosphates (Khouribga-Maroc). In: Acte du colloque International : Gestion des Rejets Industriels pour un Développement Durable (GRIDD). 1997

[21] Moufti A, Annouar S, Mountadar S, Mountadar M. The regeneration of the pre used ashes in the elimination fluorides ions from the underground waters. Journal of Materials and Environmental Science. 2016;7(6):2069-2073

[22] Annouar S, Mountada M, Soufian A, Elmidaoui A, Menkouchi Sahli MA. Defluoridation of underground water by adsorption on the chitosan and by electrodialysis. Desalination. 2004;**165**:437

[23] O.M.S. Directives de qualité de l'eau de boisson, v: 1, Recommendation. 2ème édition. Genève: Organisation mondiale de la Santé; 1994

[24] Hassani EA, Znibar A, Bouhiaoui H. Traitement du phosphate noir de Youssoufia: Amélioration du système de traitement des rejets fins et préservation de l'environnement. Annales Chimie Science Matériaux. 2001;**26**:S465-S470

Chapter 2

Greenhouse Gas Emissions of Agriculture: A Comparative Analysis

Dionisio Rodríguez

Abstract

Greenhouse gas emissions are accounted by greenhouse gases inventories, which must be produced by common accounting rules, called Guidelines, which are endorsed by the United Nations Framework Convention on Climate Change (UNFCCC). These inventories are fundamental to analyze the impact of agriculture on emissions, and as example of the difficulty and complexity of implementation of the guidelines, a comparative study is made on emissions from Agricultural Soil Management (CRF category 3D source) utilizing biological nitrogen fixation. The analysis carried out for the N_2O emissions under this section of the agrarian sector of Spain, Europe, New Zealand, Canada and the USA, inventories and national communications from Argentina and Brazil permit to observe the wide spectrum of approaches and the importance of the management of the accounting rules to be used mainly if we need that the impact of mitigation policies are captured in a direct way by the inventory. New technologies could introduce changes in the rules and can be utilized for reducing emissions, and examples are also analyzed.

Keywords: inventory of agriculture greenhouse gas emissions, N_2O emissions, biological nitrogen fixation, benchmark of countries, new technologies

1. Introduction

Agriculture is one of the economic sectors that make up the economic structure of a country and, as such economic activity, contributes to generate part of greenhouse gas of the total emissions of each country and, therefore, is an activity co-responsible for climate change.

Emissions of greenhouse gas (GHG) are accounted by greenhouse gases inventories and allow us to characterize both the emitting sources and the amount emitted and must be made respecting common rules designed with high technical qualifications.

This accounting of emissions from the agricultural sector is particularly complex and should be a useful tool for the design of agricultural policies for emissions mitigation from this sector.

To be able to check the difficulty and complexity of application of accounting guidelines and, also, the wide spectrum of options that you can use, a comparative study of the treatment of emissions from a series of inventories or national communications from various countries is made in this chapter.

It is undeniable that knowledge and the correct accounting of emissions we will strongly condition analyses and measures that specifically we design and pretend to implement to achieve lower emissions in this sector. It will enable us and also defines new technologies that may be incorporating gradually and its effect can be captured by each country GHG inventory.

2. Greenhouse inventories and agricultural sector

From the year 2015, national GHG inventories have been developed following the Guidelines of the International Panel of Climate Change (IPCC) 2006 [4]. Until that year, 1996 IPCC Guidelines were used, and the introduction of these new guidelines meant significant changes in the accounting emissions of agricultural sector. In addition to new accounting rules that affect every economic sector, the potential of global warming greenhouse gases also changed, which meant to make updates of all the data series that are measured from the 1990 base year.

Table 1 presents the global warming potential (GWP) of the three major gases that have been used and the new planned reform [1].

In addition to the changes made to the potential of global warming gases, two of which, (CH_4 and N_2O) particularly affect agriculture also changed certain accounting rules which generated significant changes in the volume of emissions in this economic sector. The changes that are made to inventories' rules will affect in proportion to each country's productive structure.

2.1 Emissions from agriculture

Emissions from agriculture activity vary depending on the economic structure of countries and the extent of its territory because agriculture is mainly based on Earth's surface arable in each country. An idea of absolute importance (total emissions) and relative (percentage of agriculture with respect to the total emissions) can be seen in **Table 2** using data on inventories [2] and national communications of different countries [3].

We can see that the developed countries have a much less percentage of agrarian sector emissions (their emissions more importantly tend to belong to the energy sector), and the big countries such as Argentina and Brazil have a large amount of emissions in relative and absolute value. An exception is New Zealand that even being a developed country has a broad agricultural sector.

Gases	Chemical formula	GWP values for 100-year time horizon		
		Second Assessment Report (SAR)	Fourth Assessment Report (AR4)	Fifth Assessment Report (AR5)
Carbon dioxide	CO_2	1	1	1
Methane	CH_4	21	25	28
Nitrous oxide	N_2O	310	298	265

Table 1.
Global warming potential (GWP) values relative to CO_2.

Year/ methodology	Countries	Agriculture emissions Total kt CO_2 eq.	Agriculture emissions/total emissions (%)
2016/AR4	EUR28 + Island	431,000	9.2
2016/AR4	USA	562,600	8.6
2016/AR4	Canada	72,000	10
2016/AR4	New Zealand	38,727	49.2
2016/AR4	Spain	34,405	12.1
2012/SAR	Argentina	119,498	27.8
2010/SAR	Brazil	407,067	32

Table 2.
Agriculture emissions by countries.

3. A comparative analysis of methodologies

In order to show the complexity and diversity of options that can be used, a comparative analysis of the methodologies used will be carried out to evaluate the emissions from agriculture's sector in the inventories and national communications of previous countries.

In a first epigraph, it will use a section of the inventory of emissions from agriculture, the emissions from managed agricultural soils: Agricultural Soil Management by the analysis that the new guidelines have been given to the nitrogen biological fixation (NBF).

We will study methodologies, which not being developed in the United Nations Framework Convention on Climate Change (UNFCCC) guidelines are approved by the inspectors in different reviews carried out inventories and if it be known by other countries could be used.

A detailed analysis is performed, with respect to N_2O emissions under this epigraph of the agrarian sector, of Spain, Europe, New Zealand, Canada and the USA inventories and national communications from Argentina and Brazil to see the broad spectrum of approaches and the importance of management of accounting rules.

We will use, as demonstrative example, legumes' crops because they are widely cultivated throughout the world on a large part of the Earth's surface, and also they have the ability to fix atmospheric nitrogen to facilitate its growth; so it is very important to know the accounting treatment that has been given to these plant species in the IPCC guidelines.

3.1 The treatment of legumes

New 2006 Guidelines introduced a very significant change in the treatment of legumes in GHG's inventories compared to previous 1996 guidelines [4]. This change meant a large reduction in emissions of the main producer countries, and this reduction is not due to an effective policy of mitigation emissions, but it is due to a simple change in accounting criteria, which is based on a technical scientist analysis.

To facilitate understanding of the problem, a brief theoretical nitrogen cycle exposure begins.

We will use the Centro Superior de Investigaciones Científicas (Spain) work, which explains very clearly this topic [5]:

> *"Nitrogen fixation may be purely abiotic or biological. In abiotic fixation, oxides are formed as a result of the combustion of organic compounds, electric shock, etc., which are dragged to the ground by rain, or ammonium by the industrial process Haber-Bosch. Biological nitrogen fixation process are carried out by prokaryotic organisms, N_2 is reduced to ammonium and incorporated into the biosphere."*

"These bacteria from the soil that we could call fixation free, as those of the genus *Azotobacter*, require up to 100 units of glucose equivalents per unit of nitrogen fixed. For this reason, its agricultural significance is low, which increases considerably in the case of the symbiotic fixation, as the established between rhizobia and legumes, where the ratio decreases of 6–12 units of glucose consumed per unit of nitrogen fixed. In this case, moreover, the power source is carbon compounds supplied directly by the plant derived from photosynthesis, while free fixation has to take them from soil, where these carbon compounds (glucose) do not exist in the amount and form necessaries. So in fact, *Azotobacter* provides to the ground a few hundred grams of nitrogen per hectare/year, and on the other hand, this value goes up in the *Rhizobium* association with alfalfa, clover, peas or soybean, until a few hundred kilos. Despite these differences, free fixation alone represents, at global level, rather less than half of the total of N_2 fixed per year [6], because symbiotic fixation, although was more high, is limited to a few plant species, including legumes." Therefore, the N_2 is fixed not only by bacteria in the roots, mostly legumes, but also by the free bacteria (not symbiotic) in the soil.

Data show that 250 Mt. of N_2 are fixed annually for bacteria and about 70 Mt. would be fixed by soil or free bacteria, which would represent 28% of N_2 fixed and about 50% would be fixed for biological fixation.

It is very important to bear in mind this data because it will strongly affect the inventories and the ways of accounting for the whole issue of the fixation of atmospheric N_2 as we will then develop.

On the one hand, it is very common that mitigation measures to tackle climate change are based on the property of legumes to fix atmospheric N_2 by a series of bacteria (genus *Rhizobium* mainly).

Thus, for example, the road map of Spain for the reduction of diffuse emissions proposes, among others, the following course of action [7]: the introduction of legumes in managed grasslands with the aim of reducing the emissions from soils in meadows. The fixation of atmospheric nitrogen produced by legumes outweighs the need for mineral fertilizers.

On the other hand, the United States of America is the only country that counts in their inventory emissions of N_2O due at atmospheric N_2 fixed by the free soil bacteria.

3.2 The measurement of the biological fixation by legumes

With the 1996 IPCC Guidelines [8] to account for emissions of nitrous oxide (N_2O) that occurs naturally in soils: "some agricultural activities bring nitrogen to the soil, increasing the amount of nitrogen (N) available for nitrification and denitrification and, ultimately, the amount of N_2O emitted. Direct emissions of N_2O from agricultural soils due to the application of N and other farming practices should reflect the contributions of anthropogenic (N) resulting from the use of synthetic fertilizers (NSF) and the animal manure applied (AMA), N of fixing varieties (NBF), the incorporation to the soils the crop residues, the nitrogen mineralization of the soil due to the cultivation of organic soils (i.e., histosols) (COS)."

The first conclusion we get is that those 1996 Guidelines address the plantation of legumes as an incorporation of N to the soil and, therefore, the producer of N_2O

emissions. The captured N of the atmosphere as a sink is not considered, but it is a source of emission.

To see how the emission due to the nitrogen biological fixation (NBF) is accounted, it can be seen from the following **Table 3** [9] that shows data of NBF in 2012 from the inventory of 15 European countries.

We can see that because the property of N fixation of legumes Europe-15 have been issued 753,000 tons of nitrogen, which then result in N_2O emissions. Transformed into CO_2 eq., they are equivalent to 4.575 Mt. of CO_2 eq.

3.3 New rules of measurement of biological fixation

As indicated above, from the year 2015, the 1996 IPCC guidelines are no longer used for inventory and entered into force the new 2006 Guidelines currently in the process of improvement.

These Guidelines say: "Biological nitrogen fixation has been removed as a direct source of N_2O because of the lack of evidence of significant emissions arising from the fixation process itself [33]. These authors concluded that the N_2O emissions induced by the growth of legume crops/forages may be estimated solely as a function of the above-ground and below-ground nitrogen inputs from crop/forage residue (the nitrogen residue from forages is only accounted for during pasture renewal). Conversely, the release of N by mineralization of soil organic matter as a result of change of land use or management is now included as an additional source. These are significant adjustments to the methodology previously described in the 1996 IPCC Guidelines."

This change means that they are accounted only for emissions from biological fixation of nitrogen for the purpose of the N_2, which are produced from the crop

Member states 2012	N-fixing crops (Gg N)
Austria	23
Belgium	4
Denmark	42
Finland	0.7
France	224
Germany	78
Greece	0.8
Ireland	0.5
Italy	140
Luxemburg	0.1
The Netherlands	4
Portugal	10
Spain	172
Sweden	35
The United Kingdom	19
EU-15	753

Table 3.
The European Union greenhouse gas inventory 2014.

residue and mineralization and so the inventories will not reflect emissions that are previously counted as biological fixation.

The amounts saved because of this new methodology are significant because they can reduce emissions under this epigraph of the inventory by 50%.

As example, if we analyze successive inventories of Spain since the year 2012–2016, we obtain the following results (**Table 4**). The 1996 IPCC guidelines were used in the year 2012 and, therefore, included the biological fixation of nitrogen emission and also applied the N_2O (SAR = 310) global warming potentials. That year was an emission result of the epigraph of agricultural soils 3D = 18,167 kt of CO_2 eq.

Subsequently, this year 2012, inventories were calculated with the new warming potential of N_2O (AR4 = 298) and began to gradually introduce the 2006 Guidelines, because as we have said is required to recalculate since 1990 with the new parameters. The result has been that the 2012 emissions calculated in the year 2018 and referred to the year 2016 have meant 9245 kt CO_2 eq. for the year 2012.

This has meant that without changing the variables of activity of this section due to the recalculations, the year 2012 emissions were almost lower 50%. The inventory lowered emissions due to a change in accounting criteria, not the implementation of mitigation measures.

Table 4 presents the evolution of Spain emissions for 2012, taking into account the recalculations marked in each annual inventory for the indicated methodological changes.

3.3.1 The cultivation of soybean

Then, we analyze the accounting treatment that different inventories of large producers of legumes make use of nitrogen biological fixation. We utilize, as an example, the crop of soybean, because it is the most widely legume cultivated worldwide and its high impact will allow us to better appreciate the distortions that occur in the accounting treatment of biological fixation. In **Table 5**, we can observe the increase of soybean crop between 2020 and 2016 surfaces mainly in Brazil and Argentina, and as in the United States, it has not changed, but remains as the maximum world producer of this crop.

These data provide us with an idea of the magnitude of the cultivation of these countries, some of which possess more hectares dedicated only to the soybean crop that the entire surface of Spain dedicated to all crops (Spain = 26.6 Million ha in the year 2016).

We can see that with such immense extensions dedicated to the cultivation of this legume, "accounting" treatment that Guidelines gives to the biological fixation will be a great importance for the inventory of emissions of these countries.

We will study two of the major producer countries (Argentina and Brazil) through their national communications and the inventories from the USA and Canada to observe how this phenomenon of biological fixation for the purposes of accounting has been treated.

Emissions of Spain in 2012 in the successive inventories in kt. CO_2				
2012	2013	2014	2015	2016
18,167	16,151	11,872	8823	9245

Table 4.
Spain emissions N_2O activity 3D.

	Soybean crop			
Year	**2000**		**2016**	
Countries	Million ha	Million tonnes	Million ha	Million tonnes
Argentina	8.6	20.1	19.5	58.8
Brazil	13.6	32.8	33.2	96.3
Canada	1.06	2.7	2.2	5.8
USA	29.3	75.1	33.5	117.2

Table 5.
Data from FAOSTAT [10].

3.3.1.1 Argentina

To analyze how this item is addressed in Argentina, in your inventory, we will use data available from the second [11] and the third national communications [12]. These two communications are still made according to the 1996 IPCC guidelines since it closes its data in 2012.

The second national communication of Argentina says "the amount of nitrogen incorporated by NBF increased around 63% between 1990/91 and 2000/01 campaign. This fact was due to the strong increase in soybean production that went from 12 to nearly 20 million tons, making it the main crop of the country. The main increase in the amount of N was due, again, to the great increase of soybean production, the main crop with contribution of the NBF."

The third communication data already indicate a rise of 5.354 Mt. CO_2 eq., that is, an increase of 31% in emissions in those 12 years (**Table 6**).

The data, in **Table 5**, show that soybean crop in Argentina increases until 19.5 Million ha and 58.8 Million tonnes in the year 2016. These data lead us to conclude that emissions from this crop and the NBF would grow strongly by applying the 1996 Guidelines. When they are finally implementing the 2006 Guidelines, the emissions reduction, due to this item, will be very significant.

As it is indicated above, Argentina will reduce its emissions simply by a change in accounting criteria unless it really is because of a mitigation policy.

The data show that the year 2012 Europe-15 saved with the accounting change of the NBF = 4.7 Mt. CO_2 eq. front the 22.5 Mt. CO_2 eq. that will save Argentina. That year 2012, Argentina only accounted 7.04 Mt. CO_2 eq. (much lower than that of the NBF) due to the use of synthetic fertilizers that are usually the most important epigraph in agriculture emissions of developed countries.

We can see that the two large producers of soybean as Brazil (33.2 Mhas.) and the USA with (33.5 Mhas.) in 2016, compared the 17.6 Mhas. of Argentina in 2012, if they counted with this methodology that would have a great impact on their emissions.

Argentina	**Years**	
	2000	**2012**
Surface of soybean (Mhas.)	8.6	17.6
Direct emissions from crops fixing (Mt. of CO_2 eq.)	17.231	22.585

Table 6.
Argentina emissions.

The Argentine Government makes a [13] comparative study to analyze the impact that would have to apply the new accounting standards (1996 Guidelines against the 2006 Guidelines) and also apply the changes of global warming potentials from methane and nitrous oxide with a result of = −58.257 Mt. CO_2 eq. if the new guidelines had been applied in the 2012 inventory (**Table 7**).

3.3.1.2 Brazil

To analyze the accounting treatment of the emissions from this crop in Brazil, third national communication [14] sent to the UNFCCC in April 2016 will be used. In this document, Brazil already uses the IPCC 2006 Guidelines and, therefore, does not consider NBF as a source of N_2O.

They also used a "study of Cardoso et al. (2008) that would demonstrate that don't exist any differences between the emissions of N_2O measured in soils planted with inoculated varieties (in Brazil, soybean is inoculated with the specific bacteria for N_2 fixation) and other varieties not inoculated." The authors of this national communication don't take into account, therefore the NBF, and they use the methodology of the 2006 Guidelines for analyses the N, which is incorporated into the soil by residue. To estimate these emissions of residues, annual productions and the amount of dry matter by crop type were used.

Brazil introduced an innovation by including a new measuring method and it explains their results also using the potential of global temperature that is proposed by the IPCC. To explain its emissions, results using three warming potential in addition to those specified in **Table 1** were used, the potential SAR and the AR5, and a new one is introduced: the global potential temperature (GTP) (**Table 8**). But, for the calculations, do not use the AR4 potential, which should be used according to the rules of implementation of the guidelines for 2006. Therefore, the results would be questionable and are not comparable.

Brazil with much more surface dedicated to the cultivation of soybean (33.2 Mhas. 2016) should produce much higher emissions than Argentina with much less soybean surface (in 2012 with 17.6 Mhas. produced 22.6 Mt. of CO_2 eq.), but the use of 2006 Guidelines that do not account NBF and the use of different potential warming involve that this country has fewer emissions.

The gap would be much better if these divergent rules had been applied to the year 2016, in which both countries doubled their crop surfaces. The conclusion is that accounting rules are very important, because Brazil "save" important

	Emissions from agriculture in Argentina in 2012 according to the different guidelines. Mt. CO_2 eq.			
Category	**IPCC 2006**	**IPCC 1996**	**Absolute difference**	**Difference %**
4A. Livestock	52.900	49.372	3.528	+7%
4.B. Agriculture	35.242	70.130	−34.887	−50%
5.A. Change of land use and forestry	63.616	90.515	−26.898	−30%
			−58.257	

Table 7.
Evaluation guidelines effect.

Gases	Chemical formula	Global temperature potential: GTP/100 years
Carbon dioxide	CO_2	1
Methane	CH_4	4
Nitrous oxide	N_2O	234

Table 8.
Global temperature potential

emissions, using the guidelines for 2006, which would be much greater with the 1996 Guidelines.

Knowing in depth how inventories are produced is, therefore, very important, and thus, we would conclude that the reduction of emissions from the agriculture sector was due to an "effective mitigation policy" when actually is due to a simple change of the accounting rules.

3.3.2 Canada

Let us look at another example, the case of Canada [15–17], which also uses the 2006 Guidelines and it is another large producer of soybeans, although in smaller amounts (5, 8 Mt. in 2016). Canada, in its 2016 inventory, reports the emissions of leguminous crops in residues which are incorporated into the soil (6.5 Mt. CO_2 eq.) [18].

As shown in **Table 9**, another novelty introduced in 2004 Canada's inventory is the appearance of new emissions due to summer fallow (0.43 Mt. CO_2 eq.) and a sink effect (−0.63 Mt. CO_2 eq.) due to the use of practices that do not till the soil, known as conservation agriculture.

These emissions and these sinks have no specific methodology in the 2006 Guidelines and their effects are calculated with methodologies developed by the country. According to the 2016 Canada's inventory, these emissions accounted for 0.22 Mt. CO_2 eq. in the case of the fallow summer and −1.5 Mt. CO_2 eq. as a sink due to conservation agriculture. Canada's inventory reports 20 Mt. CO_2 eq. as direct sources of agricultural soils (N_2O) and, therefore, conservation agriculture has been a significant sink (7,5%).

Agricultural soils (N_2O) Direct sources. Kt. CO_2 eq.	Inventory 2003	Inventory 2004	Inventory 2016
Synthetic nitrogen fertilizers	8816.77	5800	11,000
Manure applied as fertilizers	3280.76	2100	2100
Biological nitrogen fixation	3779.12	—	—
Crop residue decomposition	6154.48	3800	6500
Cultivation of organic soils	61.01	60	60
Grazing animals (pasture, range and paddock manure)	3272.71	4300	210
Mineralization of soil organic carbon	—	—	800
Conservation tillage practices		−630	−1500
Summer fallow		430	220
Irrigation			330

Table 9.
Evolution of Canada's inventories.

They also incorporate emissions from irrigation practices by which we can conclude that a great evolution has suffered the emissions of this section of direct sources of agricultural soils.

3.3.3 The United States of America (USA)

In addition to changes in the measurement of nitrogen biological fixation, the introduction of emissions' new categories, as we saw in Canada, "innovations" will continue to generate and, thus, the inventory of the United States [19], extends the computation of their emissions. This inventory includes in the accounting the nitrogen fixed, by what we have called-fixing bacteria N_2 of the soil, which belong, for example, the genus *Azotobacter* and so-called biological fixation free. The USA in your inventory calls this fixation as asymbiotic fixation and, therefore, reports emissions. (**Figure 1**) shows the scheme how are calculated agricultural soils N_2O emissions' in United States inventory.

The US inventory defines asymbiotic fixation as the fixation of atmospheric N_2 by bacteria living in the soil and that do not have direct relationship with plants. This inventory says that although the nitrogen incorporated by asymbiotic N fixation is not specifically collected by the 2006 Guidelines, it is a

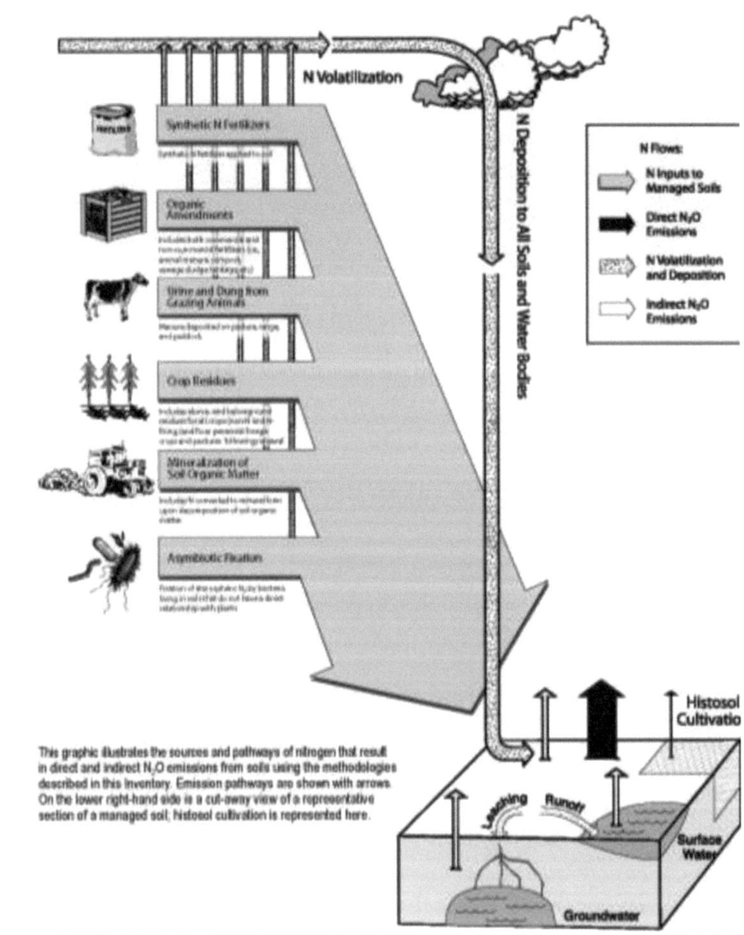

Figure 1.
Asymbiotic fixation [19].

component of the total emissions for managed soils and it should be included. It is calculated by a method of the high level developed to assess the source. To make the calculations, they use a combination of different methods using a specific model called Daycent.

The result for the year 2016 was 95.1 Mt. CO_2 eq., which includes the mineralization and asymbiotic fixation, and we can observe that it is a very significant amount. It is difficult to obtain which data belong to asymbiotic fixation and which are due to mineralization.

But what is striking is that it introduces a concept [20] that does not force the 2006 Guidelines.

We can observe in his inventory data that how meadows and crops emit more due to mineralization and asymbiotic fixation (95.1 Mt. of CO_2 eq.) by the addition of the synthetic fertilizer (64.5 Mt. CO_2 eq.) to the soils.

Only these emissions be over all Canada emissions from agricultural soils (24 Mt. of CO_2 eq.). Compared to Spain' emissions, only this item exceeds the emissions of all Spanish agriculture (34.4 Mt. CO_2 eq. in 2016) (the United States has 405 Mhas. agricultural soils and Spain 26.6 Mhas. in 2016).

We can conclude that the methodology used to measure the nitrogen biological fixation is very relevant for the emissions of a country and consideration of asymbiotic fixation (free or mineral) is an issue to consider. In the case of large agrarian countries, it would have important effects on the emission amount.

4. New technologies

In this section, we will discuss, using New Zealand GHG inventories as example [21, 22], how we can make improvements in GHG inventories and we introduce measurement methodologies of new technologies applied in the agricultural sector.

4.1 The nitrification inhibitors

The application of nitrogen fertilizers to the soil means the occurrence of biological and physicochemical reactions that leads to loss of nitrogen. The use of fertilizers with nitrification inhibitors has become a useful tool to reduce loss and improve the efficiency of the N. The use of nitrogen fertilizers stabilized become widespread and its are added, during the production process, with some substances, such as nitrification inhibitors, which can keep N applied as NH_4^+ for longer.

These products delayed the transformation of ammonia nitrogen (NH_4^+) to nitrate nitrogen (NO_3-) through temporary inhibition of various bacteria *Nitrosomonas* spp., and thus, the nitrogen is released in a progressive and gradual way and, at that same rate, it is assimilated by the crop.

Nitrification inhibitors degrade over time after being applied on the ground, and this degradation is influenced by temperature, moisture, pH and quantity of organic matter. There is already a long list of chemical compounds that have been tested as inhibitors of nitrification in the world (more than 64), but the most studied and used nitrification inhibitors are nitrapyrin, dicyandiamide (DCD) and 3,4-dimethylpyrazol phosphate (DMPP) [23].

In the United States, nitrapyrin is being used in corn, sorghum, wheat, cotton and strawberries (in a manner restricted in these). However, more than 90% is used in corn. Nitrapyrin must be injected and immediately incorporated into the soil due to its volatility and therefore its use is limited in the regions where N is typically injected to the ground. In principle, only it is marketed in the United States.

The dicyandiamide (DCD) has a bacteriostatic effect on bacteria *Nitrosomonas* spp., which only has a depressive effect on those, without killing them (not bactericidal). The disadvantage is that it requires a large amount of DCD for to contribute with between a 10% to 15% of the N-NH_4^+ to the ground. (Applications are approximately 10 kg per hectare, twice a year, in spring and autumn).

It is a very soluble product that easily seeps with rainfall separating the fraction of ammonium. Another disadvantage is that this molecule can be absorbed by the plant, and in some cases, has generated toxicity. Currently, this product is not only used but also is formulated in combination with other molecules.

3,4-dimethylpyrazol phosphate (DMPP), equally to the DCD, has a bacteriostatic effect, not the bactericidal effect (does not kill bacteria but it inhibits its action for a certain period of time), and it is relatively immobile into the soil; so it does not occur losses by leaching. On the other hand, application rates are very low compared to other nitrification inhibitors (+ −1% of the N-NH_4^+). Their application rate is 16 times lower than the rate of application of the DCD. It has a high selectivity, because it effectively inhibits only the action of *Nitrosomonas* bacteria and it degrades completely into the soil without leaving any residue. To retard the passage of ammonium to nitrate, avoiding nitrogen losses by leaching, it also reduces the effect of soil acidification.

Used as an inhibitor of nitrification, 3,4- dimethylpyrazol phosphate is regulated in the European countries and also fertilizers are used with this product in Asia and Latin America. In contrast to the DCD, in which several authors have cited toxic effects, the 3,4-dimethylpyrazol phosphate has not been demonstrated, for the moment, toxic effects on the plants.

The GHG inventory from New Zealand in the year 2012 has incorporated an amendment to the IPCC methodology that consists in introducing the use of inhibitors of the nitrification for mitigation of emissions of N_2O. They developed a methodology for incorporating the inhibitor of nitrification dicyandiamide (DCD) in the agriculture sector. N_2O emissions in the agricultural soils category take into account the use of nitrification inhibitors on dairy farms.

Based on several investigations, they have produced a good management practice that consists of the incorporation of the DCD to pastures and maximize reductions of N_2O emissions. The utilization of DCD has been reflected on the accounting and, so, incorporated in the inventory calculations and they modified parameter FracLEACH [24] and emission factor EF3PR & P [25] that are minor when using nitrification inhibitors. With these new emission factors, significant reductions of N_2O emissions from soils in both direct (nitrate leaching) and indirect (volatilization of N_2O) are achieved.

The emission factors are fixed by the Guidelines, but it is possible to modify the amount with scientific studies and this practice is done in New Zealand.

Table 10, [26] shows the differences between emission factors when DCD is not used (for example in 2012, EF3PR & P = 0.00994 front EF3PR & P = 0.01, that it is the amount fixed in the Guidelines, and FracLEACH = 0.06964 front FracLEACH = 0.07, that it is the amount fixed in the Guidelines) and as such these small differences of the emission factors meant, in total, that in 2102 its "save" 19.6 Mt. CO_2 eq.

Currently, the dicyandiamide (DCD) retired voluntarily in New Zealand's market due to the concern of customers by the existence of certain residues in dairy products even though it is at a very low level. On this point, the inventory of Agriculture of New Zealand says: "there is no risk in dairy products for humans with low levels of inhibitor used." However, in the last inventory of 2018, they asserted that sales of this product have been suspended and they have not returned to use this discount since the year 2012.

	2007	2008	2009	2010	2011	2012
Percentage of dairy area applied with inhibitor	3.5	4.5	3.1	2.2	3.0	2.9
Final modified emission factor or parameter, EF3(PRP) (kg N_2O-N/kg N)	0.00992	0.00990	0.00993	0.00995	0.00993	0.00994
Final modified emission factor or parameter, FracLEACH (kg N_2O-N/kg N)	0.06957	0.06944	0.06962	0.06973	0.06963	0.06964
Mitigation (Gg CO_2 eq.)	18.7	25.4	18.3	13.7	19.5	19.6

Note: EF3(PRP) = 0.01 and FracLEACH = 0.07 when inhibitor is not applied. All other emission factors and parameters relating to animal excreta and fertilizer use (FracGASM, FracGASF, EF4 and EF5) remain unchanged when the inhibitor is used as an N_2O mitigation technology.

Table 10.
Emission factors, parameters and mitigation for New Zealand's DCD inhibitor calculations from 2007 to 2012.

Within the existing inhibitors, 3,4-dimethylpyrazol phosphate is the inhibitor which has major advantages over the rest of nitrification inhibitors existing, due to his effectiveness at low concentrations, its stability and movement on the ground. The DMPP is an inhibitor of nitrification considerate under different national regulations of fertilizers, including the Spanish. In particular, the Royal Decree 824/2005 on fertilizer products includes fertilizers with DMPP as suitable for marketing in Spain. Similarly, Portugal has authorized the commercialization of fertilizers with DMPP.

4.2 Urea inhibitors

The New Zealand GHG inventory from agriculture has also developed a methodology for urea inhibitor called urease. Urea is the nitrogen fertilizer most used in the grasslands that are grazed in New Zealand and in addition to be excreted in the urine while the animals are grazing.

Urea inhibitors suspend or delay, during a period of time, the transformation of nitrogen in form of amide that exists in the urea to the ammonium NH_4^+ by the hydrolytic action of the urease enzyme. It reduces the speed at which urea is hydrolyzed in the soil and, therefore, losses of ammonium in the atmosphere by volatilization or nitrate by runoff are reduced or avoided.

The objective is to increase the efficiency of fertilizations with urea and to minimize the environmental impact of their use. For the purpose of the inclusion in the GHG inventory, they change the value of FracGASF [27] parameter when using the urease inhibitor.

Field- and laboratory-based studies [28] have come to the conclusion that using these inhibitors could lower FracGASF = 0.1, which is the amount fixed in the Guidelines, to a new FracGASF = 0.055, when they apply the urea inhibitor at 0.025% rates.

As a result of these practices, we can see in **Table 11** [29] a strong increase in the use of inhibitor every year and this practice has meant that in 2016 will save 20.1 kt CO_2 eq. of emissions.

Year	Percentage of urea applied that included urease inhibitor (urea treated/total urea)	Estimated greenhouse gas mitigation from using urease inhibitor kt. CO_2 eq.
2007	5.0	3.0
2008	5.2	3.0
2009	9.4	4.7
2010	6.9	4.1
2011	5.3	3.5
2012	7.0	4.6
2013	8.6	5.9
2014	20.2	13.6
2015	16.2	13.1
2016	26.5	20.1

Table 11.
Mitigation impact of urease inhibitors on nitrous oxide emissions from volatilization, from 2007 to 2016.

We can conclude that nitrification inhibitors and urea inhibitors are chemical compounds whose use can be a valid methodology to reduce the accumulation of nitrates in the soil and prevent emissions both by leaching and by volatilization of N_2O.

Different studies on mitigation policies propose the use of these practices with fertilizers or urea with inhibitors as, for example, France [30], that in a study of the INRA, proposes this mitigation measure in its roadmap for the agricultural sector. Similarly, the FAO [31], in his study of mitigation of emissions from production livestock, proposes the addiction of these inhibitors to the manure.

The use of inhibitors can be a useful tool to improve the efficiency of N in the soil, and for this reason, the use is being increased. However, this use needs still more securities, in particular, with regard to its possible effect on the food chain and in the environment and more research for the security on these products should be carried out.

Currently, these products accounted for GHG inventories do not appear and, the development of a methodology similar to the made by New Zealand could be used by other countries to reduce emissions. The 2018 USA inventory indicates that it will develop a methodology for use in next inventory due to the use of these products in the country.

4.3 The inoculation of nitrogen-fixing bacteria

The importance of legumes in the agricultural crops and its property of symbiotic fixation open the possibility of extend this property to other plant species of agricultural interest. The consequent descent of the need to use nitrogen fertilizers has made nitrogen biological fixation a subject of intense research over the years.

We will use other works of the Centro Superior de Investigaciones Científicas (Spain) for explaining this topic [32].

> *"NBF is capable of providing between 25 and 84% of the nitrogen required for normal growth and development of the cultivation of soybean. Therefore, nitrogen fixation presents great economic and ecological interest. In fact, and as an example, the high productions of soybeans around the world are due to this process through the application of microbial inoculants."*

One of the new technologies to be applied in the agricultural field that could be used for reducing emissions, in addition to other benefits, is known as biofertilization, which continues influencing mutualistic symbiosis of nitrogen fixation.

Biofertilization is defined as the use of living organisms to improve the growth of plants by two ways or increasing their nutrition making available available the required nutrients or acting on its development by the production of phytohormones. Also we can use biological control and biological remediation when with the inoculation of microorganisms we want to remove pathogens or increase the defensive response of the plants or remove xenobiotics compounds from the environment.

The use of inoculants for legumes is essential if the vegetal species has not been grown in that soil and, therefore, there is no presence of the corresponding *Rhizobium* species. This is the case of soybean in Europe that has to be inoculated with the bacteria *Bradyrhizobium japonicum* or *Sinorhizobium fredii*. This practice is also made in Brazil because most of the farmland comes from deforestation of the jungle and soybeans should be inoculated for their cultivation.

In a soil where is planted a new vegetal specie, if the natural infection's plant with these bacteria isn't possible, the crop efficiency is not superior to 40% potential. With the inoculation, up to 80% can be reached.

For example, in Spain, there are, in our soil, bacteria appropriate for crops how, alfalfa, clover, pea, lentil, chickpea, etc., but are not always effective enough; these are nitrogen fixative, and there are cases in which the inoculation is necessary if we want to obtain satisfactory yields. The same occurs when characteristics of the soil, such as acidity, drought, etc., influence the persistence of *Rhizobium* bacteria.

New knowledge is being developed in this field of biological nitrogen fixation and investigators start to get the extension of the fixing capacity to other non-leguminous plants of interest. So, they are trying to achieve that corn, wheat or rice be infected efficiently by *Rhizobium*, and begin to glimpse the possibility of transferring to these plants the fixing capacity. So its plants will be able to take advantage of the atmospheric N_2 for themselves.

In the short term, the selection of strains and their appropriate genetic manipulation are underway, to prepare the most suitable inoculants and, also, to improve the plant so that there are no limiting factors in the establishment of the corresponding symbiosis.

> *"Researchers point that not everything is optimal in obtaining self-sufficient plants for nitrogen, because although the crops should not be fertilized, it would be less productive. The energy cost involved in fixing becomes up to three times higher than the utilization of nitrate and the plants would grow less, the performance would be lower and may even were reduced the area of cultivation. But this independence of nitrogen fertilization would possibly more profitable crop, more suitable for economically weak areas and environmentally cleaner" [32].*

In short, as we have seen above, the 2006 Guidelines do not consider the contribution of nitrogen by biological fixation, which involves a direct emission of N_2O to the atmosphere and, therefore, this technology should be taken into account.

5. Conclusions

The GHG inventory is a great source of information, not only for its environmental aspect, but also by the possibility of using their data for relevant technical and economic analyses. Other quality is its role for serving us a guidance when preparing mitigation policies.

This comparative study of different inventories show the wide spectrum of approaches and the importance of the management of the accounting rules.

A detailed analysis of the nitrogen biological fixation and, particularly, the cultivation of soybeans, allow us to appreciate the importance of the follow-up to the guidelines that govern the preparation of inventories.

This article also shows significant differences in the volume of emissions due to the use of the 1996 Guidelines front the 2006 Guidelines, and both change the rules and the changes of global warming potentials. Using the emissions from the cultivation of an important legume such as Argentina and Brazil soybeans, we can observe in a practical way the importance of methodological changes in "accounting standards."

The case of the USA includes emissions that would not be bound by these guidelines, such as those due to asymbiotic fixation and the case of Canada incorporates the non-tillage (conservation agriculture) sink effect as the emissions due to fallow and irrigation systems.

On the other hand, examples of introduction of new technologies are exposed that are not included in the Guidelines, which require the development of specific methodologies. The case of the inventory of New Zealand regarding the nitrification inhibitors and urea inhibitor is a relevant example. Nitrogen Biological Fixation should be one of the fields to research in more depth because the specificity of some bacteria to capture atmospheric N_2 could provide large reductions in the use of synthetic nitrogen fertilizers.

Although the guidelines seek to unify criteria, we have exposed the full spectrum of options that have all these inventories to the same heading of emission from agricultural soils. We must bear in mind that correct accounting will condition strongly the analyses and measures that specifically we design and try to implement in the agricultural sector.

The objective should be that the effectiveness of a mitigation policy was validated for the concrete results of the GHG inventory and, of course, that this policy could be applied at the farm level. Deep knowledge of accounting rules is a necessary premise. It is very important that all persons participating in the measurement of emissions (technicians, researchers, public, professional managers of agriculture, etc.) are aware of the use of the same rules.

A country greenhouse gas measurement methodology is a science, not well known among professionals, which requires, in addition to a large sectoral specialization, to be addressed, by large multidisciplinary teams, and should be given in college and, perhaps, the most difficult, to be conducted at the level of agricultural farm in an understandable way. We must bear in mind that farmers will be responsible for putting into practice any measure of mitigation or generation of sinks that intends to.

Author details

Dionisio Rodríguez
Xunta de Galicia, Santiago de Compostela, Spain

Address all correspondence to: dionisio.rodriguez.alvarez@gmail.com

References

[1] IPCC Assessment Reports. Available from: http://www.ipcc.ch/publications_and_data/publications_and_data_reports.shtml

[2] National Inventories Submisions. Available from: https://unfccc.int/process-and-meetings/transparency-and-reporting/reporting-and-review-under-the-convention/greenhouse-gas-inventories-annex-i-parties/national-inventory-submissions-2018

[3] National Communication Submisions. Available from: https://unfccc.int/process#:0c4d2d14-7742-48fd-982e-d52b41b85bb0:f666393f-34f5-45d6-a44e-8d03be236927:cc852874-8331-492c-a332-cc6313dec434

[4] International Panel of Climate Change. UNEP.WMO (2006) Directrices del IPCC de 2006 para los inventarios nacionales de gases de efecto invernadero. IPCC. Available from: http://www.ipcc-nggip.iges.or.jp/public/2006gl/spanish/pdf/4_Volume4/V4_11_Ch11_N2O&CO2.pdf

[5] Pascual JO. Fijación biológica de Nitrógeno. (Última Act. 08-02-2008). CSIC, Granada: Estación Experimental del Zaidín; 2008. Available from: https://www2.eez.csic.es/olivares/ciencia/fijacion/

[6] Pascual JO. Fijación biológica de Nitrógeno. (Última Act. 08-02-2008). CSIC, Granada: Estación Experimental del Zaidín; 2008. Available from: http://www.eez.csic.es/~olivares/ciencia/fijacion/

[7] Oficina Española de Cambio Climático. Hoja de ruta de los sectores difusos a 2020. MAGRAMA. Oficina Española de Cambio Climático; 2014

[8] Revised 1996 IPCC guidelines for National Greenhouse Gas Inventories. 1996. Available from: https://www.ipcc-nggip.iges.or.jp/public/gl/invs1.html

[9] EEA, European Environment Agency. Annual European Union Greenhouse Gas Inventory 1990-2012 and Inventory Report 2014. European Environment Agency. 2014. Available from: http://www.eea.europa.eu/publications/european-union-greenhouse-gas-inventory-2014

[10] FAOSTAT. Available from: http://www.fao.org/faostat/en/#data/QC

[11] Segunda comunicación nacional de la República Argentina a la convención marco de las Naciones Unidas sobre el cambio climático. 2008. Available from: https://unfccc.int/documents/67498

[12] Tercera comunicación nacional de la República Argentina a la convención marco de las Naciones Unidas sobre el cambio climático.2015. Available from: https://unfccc.int/documents/67499

[13] Asociación Argentina de Consorcios Regionales de Experimentación Agrícola (AACREA), Fundación Torcuato Di Tella (FTDT), Price Waterhouse & Co. Asesores de Empresas SRL. Informe comparativo de las Directrices IPCC 1996 vs. 2006 Agricultura, Ganadería, y Cambio de Uso del Suelo y Silvicultura. 2015. Available from: http://ambiente.gob.ar/archivos/web/ProyTerceraCNCC/file/5_%20Comparativa%20Gu%c3%adas%20IPCC%201996%20y%202006%20-%20Agricultura,%20Ganaderia%20y%20CUSS.pdf

[14] Third National Communication of Brazil to the United Nations Framework Convention on Climate Change. Available from: https://unfccc.int/documents/66129

[15] Canada – Initial Submission 2005 GHG Inventory 1990-2003.

Available from: https://unfccc.int/process/transparency-and-reporting/reporting-and-review-under-the-convention/greenhouse-gas-inventories/submissions-of-annual-greenhouse-gas-inventories-for-2017/submissions-of-annual-ghg-inventories-2005

[16] National Inventory Report 1990-2004. Greenhouse Gas Sources and Sinks in Canada — Final Submission. Available from: https://unfccc.int/process/transparency-and-reporting/reporting-and-review-under-the-convention/greenhouse-gas-inventories/submissions-of-annual-greenhouse-gas-inventories-for-2017/submissions-of-annual-ghg-inventories-2006

[17] National Inventory Report 1990-2016. Greenhouse Gas Sources and Sinks in Canada. Available from: https://unfccc.int/documents/65715

[18] "There is no evidence that measurable amounts of N2O are produced in Canadian agricultural soils during the N fixation process itself. Therefore, Canada decided to report this source as "not occurring." However, the contribution of legume N to N2O emissions is included as a source of N2O emissions from crop residue decomposition on agricultural soils (NRES)".The Canadian Government. (2014). National Inventory report. 1990-2012. Green house gases and sources in Canada. The Canadian Government's Submission to the UN Framework Convention on Climate Change. 2014. Available from: https://unfccc.int/process-and-meetings/transparency-and-reporting/reporting-and-review-under-the-convention/greenhouse-gas-inventories-annex-i-parties/national-inventory-submissions-2018

[19] Inventory of U.S greenhouse gas emissions and sinks 1990-2016. EPA 2018. Available from: http://www.epa.gov/ghgemissions/inventory-us-greenhouse-gasemissions-and-sinks-1990-2016

[20] Inventory of U.S. Greenhouse Gas Emissions and Sinks: 1990-2012. U.S. Environmental Protection Agency. U.S.A; 2014. Available from: http://unfccc.int/files/national_reports/annex_i_ghg_inventories/national_inventories_submissions/application/zip/usa-2014-nir-15apr.zip

[21] "N inputs from asymbiotic N fixation are not directly addressed in *2006 IPCC Guidelines*, but are a component of the total emissions from managed lands and are included in the Tier 3 approach developed for this source". U.S. Environmental Protection Agency. (2014). Ministry for the Environment. New Zealand's Greenhouse Gas Inventory 1990-2012. Ministry for the Environment. 2014. Available from: http://unfccc.int/files/national_reports/annex_i_ghg_inventories/national_inventories_submissions/application/zip/nzl-2014-nir-14apr.zip

[22] Ministry for the Environment. New Zealand's Greenhouse Gas Inventory 1990-2016. Ministry for the Environment. 2018. Available from: https://unfccc.int/documents/65588

[23] Trenkel ME. Slow- and Controlled-Release and Stabilized Fertilizers: An Option for Enhancing Nutrient Use Efficiency in Agriculture. Paris, France: International Fertilizer Industry Association (IFA); 2010. Available from: https://www.google.es/?gws_rd=ssl#q=Slow-+and+Controlled-Release+and+Stabilized+Fertilizers:+An+Option+for+Enhancing+Nutrient+Use+Efficiency+in+Agriculture

[24] FracLEACH-(H) = fraction of all N added to/mineralised in managed soils in regions where leaching/runoff occurs that is lost through leaching and runoff, kg N (kg of N additions)-1

[25] EF3PRP = emission factor for N_2O emissions from urine and dung N deposited on pasture, range and paddock by grazing animals, kg $N_2O–N$ (kg N input)-1

[26] Ministry for the Environment. New Zealand's Greenhouse Gas Inventory 1990-2016. Ministry for the Environment; 2018. p. 198. Available from: https://unfccc.int/documents/65588

[27] FracGASF = fraction of synthetic fertiliser N that volatilises as NH_3 and NOx, kg N volatilised (kg of N applied)-1

[28] "Saggar et al. (2013) showed that the presently recommended country-specific value of FracGASF of 0.1 be reduced to 0.055. This finding was based on field and laboratory studies conducted both in New Zealand and worldwide." Ministry for the Environment. New Zealand's Greenhouse Gas Inventory 1990 – 2016. Ministry for the Environment; 2018. p. 196

[29] Ministry for the Environment. New Zealand's Greenhouse Gas Inventory 1990-2016. Ministry for the Environment; 2018

[30] European Parliament. Measures at farm level to reduce greenhouse gas emissions from EU agriculture. Directorate General for Internal Policies. Policy Department Structural and Cohesion Policies. Agriculture and Environmental Development. European Parliament; 2014

[31] Gerber PJ, et al. Mitigación de las emisiones de gases de efecto invernadero en la producción ganadera. Una revisión de las opciones técnicas para la reducción de las emisiones de gases diferentes al CO_2. FAO; 2013. Available from: https://www.google.es/?gws_rd=ssl#q=Mitigaci%C3%B3n+de+las+emisiones+de+gases+de+efecto+invernadero+en+la+producci%C3%B3n+ganadera.+Una+revisi%C3%B3n+de+las+opciones+t%C3%A9cnicas+para+la+reducci%C3%B3n+de+las+emisiones+de+gases+diferentes+al+CO2.+FAO+2013

[32] Pascual JO. Tiene futuro la fijación biológica de nitrógeno? CSIC, Granada: Estación Experimental del Zaidín; 2008. Available from: http://www.eez.csic.es/~olivares/ciencia/futuro2/index.html

[33] Rochette P, Janzen H. Towards a revised coefficient for estimating N_2O emissions from legumes. Nutrient Cycling in Agroecosystems. 2005;**73**(2):171-179

Chapter 3

Progressive Research in the Molecular Mechanisms of Chronic Fluorosis

Liming Shen, Chengyun Feng, Sijian Xia, Yan Wei, Hua Zhang, Danqing Zhao, Fang Yao, Xukun Liu, Yuxi Zhao and Huajie Zhang

Abstract

Long-term excessive intake of fluoride (F) leads to chronic fluorosis, resulting in dental fluorosis and skeletal fluorosis. Chronic exposure to high doses of fluoride can also cause damage to soft tissues, especially when it passes through the blood-brain, blood-testis, and blood-placenta barrier, causing damage to the corresponding tissues. Fluorosis has become a public health problem in some countries or regions around the world. Understanding the pathogenesis of fluorosis is very important. Although the exact mechanism of fluorosis has not been fully elucidated, various mechanisms of fluoride-induced toxicity have been proposed. In this chapter, we will introduce the research progress of the mechanism of fluorosis, focusing on dental fluorosis, skeletal fluorosis, nervous and reproductive system toxicity, and influential factors related to fluoride toxicity (i.e., genetic background, co-exposure with other element). In addition, the application of proteomics and metabolomics in the study of the pathogenesis of fluorosis is also introduced. Currently, there is still no specific treatment for fluorosis. However, since fluorosis is caused by excessive intake of fluoride, avoiding excessive fluoride intake is the critical measure to prevent the disease. In endemic regions, health education and supplement diet with vitamins C, D and E, and calcium and antioxidant compounds are important.

Keywords: chronic fluorosis, fluoride, influential factor, mechanisms, proteomics, skeletal fluorosis

1. Introduction

Fluorine is a highly active gaseous element found widely in nature. Fluoride in small doses is beneficial for preventing dental caries and is commonly used in the prevention of dental caries [1, 2]. However, long-term excessive fluoride intake will affect human health, causing chronic fluorosis. Chronic fluorosis is a systemic disease, high doses of fluoride leads to bioaccumulation in the body, especially hard tissues such as bones and teeth, and primarily harms bones and teeth [3–5]. Besides skeletal and dental damage, excessive exposure to fluoride can also cause other non-phrenological hazards, such as metabolic, structural, and functional damage to the nervous system [6–12], kidneys [13–16], liver [14, 16–19], cardiovascular system [20–23], and reproductive system [24–26].

Chronic fluorosis is an endemic disease; it is endemic in at least 25 countries across the globe, China and India being the worst affected among them [27]. Most cases of fluorosis were caused by drinking fluorous water. In China, fluorosis is caused by drinking water as well as inhaling combustion fumes of coal being used as an indoor fuel source [28–31]. Guizhou is one of the most severely afflicted areas of endemic fluorosis in China and this occurrence is due to indoor coal burning [30]. Another type of fluorosis is brick tea-type fluorosis, due to fluoride accumulation in brick tea. It is more prevalent in Tibet than in other regions of China [32]. It is also worth noting that chronic exposure to volcanic environments may lead to the exposure of excessive amounts of fluoride [33]. It is estimated that more than 10% of the worldwide population live within the potential exposure range of some active or historically active volcano, either erupting or in a post-eruption phase [34].

In recent years, numerous studies focused on the molecular mechanisms associated with fluoride toxicity [35–39]. Although the underlying mechanisms of chronic fluorosis is still not well understood, the results of the previous studies indicated that fluoride can induce oxidative stress; regulate intracellular redox homeostasis; and lead to mitochondrial damage, endoplasmic reticulum stress, and alteration of gene expression [35–39]. Other mechanisms include enzyme inhibition, induction of apoptosis, cell cycle arrest, etc. [35–39]. This chapter reviews the present research on the potential adverse effects of overdose fluoride on various organisms, summarizes the molecular mechanism of fluorosis, and aims to improve our understanding of fluoride toxicity.

2. Mechanisms of skeletal fluorosis

Fluoride is a cumulative toxin, which accumulates in mineralized tissues, notably in the lattice of bone and tooth crystals. The bones and teeth are recognized as the target organs of fluoride, and bone tends to accumulate this element with age. The main features of the disease are dental fluorosis and skeletal fluorosis. Dental fluorosis is the first visible toxic effect of F exposure, which manifests as pitting of tooth enamel and yellow cracked teeth in adults and in children [37]. Skeletal fluorosis is a metabolic bone disease with osteosclerosis as the major clinical sign, mostly involving bone joints [40]. It results in ligament calcifications, accompanied by osteopenia, osteoporosis, and osteomalacia to varying degrees [40, 41]. Fluorine is a trace element that is incorporated into bone mineral during bone formation [42]. Fluoride substitutes for the hydroxyl group in hydroxyapatite, forming fluorapatite. Bone metabolism includes the process of osteoblasts forming bone and the osteoclasts degrading bone. Fluoride has an effect on bone mineral, bone cells, and bone architecture [42]. Fluoride at physiological levels promotes osteoblast proliferation, increases bone mass, as well as increases osteoblast activity via the up-regulation of markers such as alkaline phosphatase (ALP), bone morphogenetic protein (BMP), and bone gla protein (BGP) [43]. The levels of ALP and BGP were higher in patients with skeletal fluorosis than the control group [44]. However, fluoride may stimulate osteoblastic activity and delay mineralization of new bone [42]. On the other hand, osteoclasts are derived from hematopoietic progenitors in bone marrow and are only responsible for bone resorption. The mechanism associated with the osteoclasts is complicated; some studies showed that high fluoride concentrations may promote the formation of osteoclasts [45], or reduce the number of osteoclasts and decrease their bone resorption ability [46, 47]. Others suggested that fluoride had little effect on the number of osteoclasts and no effect on the osteoclast formation

[48]. Indeed, excessive fluoride intake can destroy the processes of bone formation and resorption, which may lead to bone turnover disorders and result in skeletal fluorosis. Bone turnover is a dynamic balance regulated by osteoblasts and osteoclasts. Excessive fluoride disrupts this balance, influencing the differentiation of osteoblasts and osteoclasts and resulting in the development of bone lesions [49]. This may be related to certain signaling pathways and mechanisms. Fluoride influences bone turnover by regulating certain factors, such as runt-related transcription factor 2 (Runx2) and receptor activator for nuclear factor-κB ligand (RANKL), which act as markers of osteoblasts and osteoclasts [48, 50]. Through the mitogen-activated protein kinases (MAPK) pathway, fluoride mediates gene expression and cell viability. In ameloblasts, fluoride activates the Rho/ROCK pathway. Fluoride can also induce endoplasmic reticulum (ER) stress, leading to protein misfolding [51]. In addition, TGFβ-SMAD signaling regulates expression of essential genes (MMP13, Collagen Type I, Collagen Type VII, Aggrecan, and Biglycans) involved in the formation of extracellular matrix (ECM). Fluoride exposure affects the expression of these genes through TGFβ-SMAD signaling [52]. Additionally, excessive fluoride exposure leads to disturbances of bone homeostasis. c-Fos is known to be essential in bone development by affecting osteoblast and osteoclast differentiation, suggesting that c-Fos might negatively regulate osteoprotegerin (OPG) expression induced by fluoride in osteoblastic cells [53].

Furthermore, collagen and noncollagenous proteins are of significant importance for maintaining the biomechanical integrity of the bone and many bone matrix proteins play important roles in mineralization [42]. Excessive intake of fluoride affects the bone matrix proteins, that is, collagen and noncollagenous proteins, which may be another possible mechanism of skeletal fluorosis [42, 54]. For example, it has been shown that fluoride could inhibit the synthesis of type I collagen and decrease the degree of collagen cross-linking [54–59] or affect other collagen proteins [60–62], and affect the synthesis of proteoglycan [63], and expression of matrix metalloproteinases (MMPs) [54, 64, 65]. Taken together, these studies suggested that exposure to fluoride alters growth, ECM formation, bone mineralization, and skeletal development and induced bone formation and bone resorption, thus leading to the development of fluorosis.

3. Nervous system toxicity

Excessive fluoride may cross the blood-brain barrier and accumulate in the brain, causing dysfunction of the central nervous system (CNS). In recent years, many studies have focused on fluorine neurotoxicity. The central nervous system during development is highly sensitive to the influence of fluorine due to its weakened protective mechanisms [66]. Studies showed that children in high fluoride areas had significantly lower IQ (intelligence quotient) scores than those who lived in low fluoride areas [67, 68]. The results of meta-analyses supported the possibility of adverse effects of fluoride exposures on children's neurodevelopment [67, 69]. In the animal experiments, as exposed to high levels of fluorine, the content of fluorides in the rats' brains was even 220 and 300 times higher than in the control group [70], and fluoride exposure affects the behavior, memory, cognitive and learning ability [71–73]. Dendritic thickening and disappearance, mitochondrial swelling, neuronal endoplasmic reticulum dilation, and impaired hippocampus synaptic interface structure can be observed in the brain of fluoride exposed rats [73]. The numbers of Nissl bodies in neurons in the hippocampus and cortex of brains from both adult rats and their pups with fluorosis were reduced, suggesting an injury of neurons [10]. These data indicate that excessive exposure to fluoride results in structural and functional damages to the central nervous system, and may significantly hinder the neurodevelopment.

Fluorine neurotoxicity may be associated with oxidative stress, neuroinflammatory and neurotransmitter alterations. Fluorine induces increase in ROS (reactive oxygen species) and lipid peroxidation and decrease in anti-oxidative enzyme activity in neurons and glia, resulting in oxidative stress, which in turn causes cell damage and metabolism disorders [12, 74]. Fluorine causes glial cell activation which is involved in inflammation through producing proinflammatory cytokines. Chronic inflammation in the brain appears to cause neuronal damage [66, 75]. Moreover, fluorine influences the synthesis of neurotransmitters, the activity of enzymes, the expression of receptors, and the plasticity of neurons [76–78]. Therefore, excessive exposure to fluoride results in structural and functional damages to the central nervous system.

Of note, because fluoride can not only cross the blood-brain barrier, but also penetrate through the placenta, fluorine exposure in the prenatal and neonatal periods is dangerous [66, 79]. A recent study showed that during pregnancy and lactation, even at very low concentrations, F exposure may alter parameters of the central nervous system functionality, producing a delay in eye-opening development in the offspring as well as hypoactivity in adult offspring [80]. Further studies will be crucial to elucidate the molecular mechanisms through which F exposure during gestation and lactation trigger neurobehavioral changes [80].

4. Reproductive system toxicity

Research on the effects of fluoride on the reproductive system has been carried out for many years. As early as 1925, Schulz and Lamb reported the reproductive toxicity of fluoride [81]. Fluoride shows adverse effects on the male reproductive system, including spermatogenesis defect, sperm count loss, sperm differentiation, and maturation impairment [82], and increase in chromosomal aberrations in primary testicular cells and the rate of sperm deformity [83]. Interestingly, recent studies showed that exposure to fluoride can alter the BTB (blood-testis-barrier) [84, 85]; fluoride induced structural and functional alterations in the BTB by increasing the expression levels of Arp3 protein with a concomitant increase in the expression levels of IL-1α (interleukin-1α) that led to the reorganization of the highly branching F-actin and the decreased expression of F-actin [25]. A significant increase in the fluoride concentration in the testes of mice that were exposed to sodium fluoride (NaF) has been observed [85]. In addition, ovaries of albino rats treated with high doses of NaF exhibited abnormal ovarian follicles, dilated blood vessels, stromal congestion, and necrotic granulose cells [86].

Cell apoptosis is one early sign of genotoxic damage in mature testis, and plays critical roles in spermatozoa output. Fluoride may induce oxidative stress through the activation of MAPK cascade and Jun N-terminal kinase (JNK, c-Jun) and extracellular signal-regulated protein kinase (ERK) signaling pathway lead to cell apoptosis that includes both intrinsic and extrinsic apoptotic pathways [82]. Fluoride could also cause leakage of potassium ions, thereby reducing sodium and potassium levels in spermatozoa [87]. In addition, higher levels of inflammatory factor such as IL-1α were detected in the testes of NaF-treated rats [25], suggesting that inflammation was involved in the in the toxicity of fluoride to the reproductive system [25, 88]. More recently, a proteomics study analyzed the proteome characteristics of sperm from fluoride-exposed mice, and identified 15 differentially expressed proteins between fluoride-exposed and control groups. Most of them are associated with sperm functions such as sperm motility, maturation, capacitation and acrosome reaction, lipid peroxidation, detoxification, inflammation, and stability of membrane structure [89]. Another study reported altered MicroRNA (miRNA) expression profiling in sperm of mice induced by fluoride. Sixteen altered miRNAs were identified and they mainly were involved in protease inhibitor activity, apoptosis, ubiquitin-mediated proteolysis, and signaling pathways of calcium, JAK-STAT, MAPK, p53, and Wnt [90]. These findings provide new insights into the mechanism underlying fluoride reproductive toxicity. However, the toxicity mechanism of fluoride on the reproductive system still needs further exploration.

5. Other systems

As mentioned above, excess fluoride uptake affects other organs including liver and kidneys, and cardiovascular system. Liver is the most important detoxification organ in the body. The effect of fluoride on the liver has been widely studied and it has been demonstrated that excessive intake of fluoride causes serious liver damage [14, 16–19].

The kidneys are the main route of F removal from the body, and approximately 60% of the total daily F absorbed is filtered and excreted in urine [91]. The link between fluoride and kidney disease has been known and confirmed for many years [13–16], the toxicity or damage of fluoride to the kidney has been observed in population and experimental animals, including the kidneys of the fetus and suckling mammal [92]. Of note, people on kidney dialysis, patients with reduced glomerular filtration rates, and diabetic mammals are particularly susceptible to fluoride exposure [15].

A rising number of research studies have been carried out on the toxic effect of F in cardiovascular system [20–23, 93]. Fluoride can accumulate in the cardiovascular system, resulting in arterial calcifications, elastic properties of ascending aorta disruption, and ventricular diastolic dysfunction [93].

6. Influence factor

6.1 Genetic susceptibility to fluorosis

Clearly, toxic effects in humans due to chronic fluoride ingestion mainly depend on the total dosage and duration of exposure. However, dose and time alone are not the only factor affecting fluorosis. Some studies have shown the existence of non-responder populations to fluorine [94], while others have shown that some people seem to be very sensitive to fluorine [95, 96]. Animal experiments have observed that three inbred strains of mice (A/J, SWR/J, 129P3/J) displayed variations in the onset and severity of dental/enamel fluorosis with equivalent fluoride exposure [97]. The bone mechanical properties were reduced in the "susceptible strain" (A/J), moderately altered in the "intermediate strain" (SWR/J), and unaffected in the "resistant strain" (129P3/J), suggesting a genetic contribution to the variation in bone response to fluoride content [97]. Fluoride effects on bone formation and mineralization are influenced by genetics [42]. Another study showed that exposure to the same dosage and time, as compared with Wistar rats, the urine fluoride of SD rats was higher while bone and teeth fluoride levels were lower. Meanwhile, dental fluorosis susceptibility of SD rats is higher [98].

Interestingly, the association between genetic polymorphisms in candidate genes and the susceptibility in the development of fluorosis has been well reviewed [27]. Candidate genes associated with human dental fluorosis and skeletal fluorosis are listed in **Table 1**. Candidate genes in dental fluorosis include BGLAP (Osteocalcin), COL1A2 (Collagen type 1 alpha 2), CTR/CALCR (Calcitonin Receptor), ESR (Estrogen Receptor), and VDR (Vitamin D Receptor). Candidate genes in skeletal fluorosis include MMP-2 (Matrix metallopeptidase2), MPO (Myeloperoxidase), GSTP1 (Glutathione S-transferase pi 1), PRL (Prolactin), and VDR (Vitamin D Receptor). These genes are involved in different functions—BGLAP, ESR, and COL1A2 are related to bone formation and development; VDR and CTR are related to bone formation and metabolism; and PTH and PRL are related to hormones' secretion. GSTP1, MMP, COMT, and MPO are related to detoxifying enzymes, extracellular matrix, cognitive and immune responses, respectively [27]. These results suggest that an individual's genetic background plays a major role in influencing the risk to fluorosis.

6.2 Co-exposure with other element

Co-exposure to other elements is another major factor affecting fluorosis. All of these could complicate the overall toxic response. For example, in geothermal areas, volcanic activity includes CO_2-rich hot springs, steaming vents, hot ground and boiling mud pools that normally contain unusually high concentrations of Li, Rb, Cs, Si, B, As, and F [33]. Thus, chronic exposure to volcanic environments may lead to the exposure of excessive amounts of fluoride and other elements. Interestingly, a recent study indicated that an increase or decrease in various elements (including F, Al, Se, Zn, Cu, Fe, Mo, Mn, B, V, Ca, Mg, and P) in the environment is related to the abnormal levels of the corresponding elements in a fluoride-exposed population [28]. High levels of F, Al, As, Pb, and Cr were a risk factor for dental fluorosis, but not Se, Zn, Cu, B, Ca, and P, which was a protective factor for dental fluorosis [28].

Candidate genes	Polymorphism site (restriction sites or mutational bases)	References
Dental fluorosis		
COL1A2 (Collagen type1 alpha 2)	rs414408 (PvuII) rs412777 (A/C)	[99] [100]
ESR (Estrogen receptor)	rs1256049 (G > A, RsaI) rs2234693 (A > C, XbaI)	[101]
AMBN (Ameloblastin)	rs4694075 (C/T)	[102]
TFIP11 (Tuftelin interacting protein 11)	rs5997096 (C/T)	[102]
TUFT1 (Tuftelin)	rs4970957 (A/G)	[102]
DLX1 (Homeobox protein DLX-1)	rs788173 (A/G)	[103]
DLX2 (Homeobox protein DLX-2)	rs743605 (A/G)	[103]
TIMP1 (Metalloproteinase inhibitor 1)	rs4898 (C/T)	[103]
Skeletal fluorosis		
FRZB1 (frizzled-related protein 1)	rs2242070 (A/G)	[104]
VDR (Vitamin D receptor)	rs2228570 (Fok I)	[105]
GSTP1 (Glutathione S-transferase pi 1)	rs1695 (A/G)	[106]
PRL (Prolactin)	rs1341239	[107]
MMP-2 (Matrix metallopeptidase 2)	rs2287074 (G/A) rs243865	[108]

Table 1.
Candidate genes in dental fluorosis and skeletal fluorosis.

At present, some studies have been reported on the co-exposure of fluorine and arsenic (As) or fluorine and aluminum (Al). Both arsenic and fluoride are ubiquitous in the environment. The co-exposure of fluorine and arsenic is mainly through drinking water [109–111] or burning coal [112, 113]. The latter is a unique type in China, which was attributed to exposure to high levels of As and F in food and breathing As-laden air, caused by polluted food and air due to indoor combustion of coal [112, 113]. The interaction mechanism of these two elements is complicated, which may be independent, synergistic, or antagonistic [114]. A recent study indicated that arsenic may be involved in fluoride-induced bone toxicity through PTH/PKA/AP1 signaling pathway [115]. Arsenic affects the expression of c-Fos, thereby affecting the expression of transcription factor AP1, indirectly involved in fluoride-induced bone toxicity [115]. Another study showed that the joint effect of fluoride and arsenate on the gene expression of ODF (osteoclast differentiation factor) is antagonistic, while the combined effect on the gene expression of OPG is synergistic [116]. Ma et al. reported that As and F can induce the expression of adhesion molecules, chemokines and pro-inflammatory cytokines in rabbit aorta separately, and antagonistic effects were observed on inflammatory response [117]. Fluoride and arsenic, either alone or combined, can decrease learning and memory ability in rats [118].

Combined exposure to fluoride and aluminum is another noteworthy problem related to fluorosis. It mainly occurs through indoor combustion of coal, especially kaolin mixed with coal [119], and high Al content in tea such as brick tea [120]. The interaction mechanism of F and Al is also complicated, may be independent, synergistic, or antagonistic. Aluminum exposure impairs bone formation; inhibition of bone formation by aluminum through different signal transduction pathways has been reported [121]. Exposure to Al is associated with low bone mineral density (BMD) and an increased risk of osteoporosis [121–124]. Fluoride enhances the uptake of aluminum; the simultaneous administration of fluorine and aluminum increased plasma [125], and bone [126] concentrations of aluminum in rats, whereas aluminum suppresses the uptake of fluoride [127]. Decreased bone mineral density was observed in fluorine and aluminum-treated rats [126]. Patients with co-exposure to fluoride and aluminum display with osteomalacia or osteoporosis may be due to fluoride promoting aluminum accumulation in bone, while aluminum inhibits bone formation. However, the vitro study showed that there was a synergistic effect of fluoride and aluminum on the expression of Runx2 and Osterix mRNA in osteoblastic MC3T3-E1 cells, thereby enhance MC3T3-E1 cells proliferation and differentiation [128], and contribute to osteosclerosis. This may explain the different clinical features of skeletal fluorosis, that is, osteosclerosis accompanied with osteomalacia, and osteopenia.

It is worth mentioning that Al is a well-known neurotoxic agent, and it has long since been implicated in the etiopathology of AD [129]. Fluorine and aluminum are able to cross the blood-brain barrier and the placental barrier [66, 130]. They can accumulate in the brain, and fluoride did not affect the accumulation of aluminum in the CNS [131]. It has been reported that increases of microglia activation and inflammatory response were seen in aluminum, fluoride, and a combination of aluminum-fluoride-treated rat brain [132]. Excessive fluoride and aluminum intake induces the progression of cell death which inhibits acetylcholinesterase (AChE) activities and triggers the release of lysosomal and cell cycle proteins in the brain of rats [133]. More recently, Xie et al. found that continuous exposure to fluorine and/or aluminum of mother rats impaired the neurobehavioral reflexes, spatial learning, and memory of offspring rats [134, 135]. The effects of F were obvious, but the effects of Al were slight. There were antagonistic effects between F and Al, with Al reduction in the toxicity of F [135].

In addition, Chinoy et al. reported that simultaneous exposure of the animals to NaF and $AlCl_3$ was associated with an increased toxic effect on gonadal steroidogenesis, uterine metabolism of carbohydrates, and hypercholesterolemia, as compared with each compound administered separately [136]. Recently, Dong et al. reported that F induced the reduction in testosterone and sperm amount; however, Al had antagonism effects on F and weakened the toxicity of F to some extent [137]. Moreover, fluoride interacts with aluminum to form a fluoro-aluminum complex AlFx (e.g., AlF_3 and AlF_{4-}), which can interact with the G protein (guanine nucleotide-binding proteins) and activated effect or enzymes, providing false information, and amplify the processes of signal transmission [138]. Together, further investigation is needed on the underlying mechanisms by which fluorine and arsenic or fluorine and aluminum induce toxicity.

7. Proteomics and metabolomics applications

Proteomics and metabolomics are useful and powerful tools for clarifying toxicological mechanisms associated with diseases. Proteomics offers the possibility to map the entire proteome of an organism or cells and detect toxic effects at significantly lower doses, as well as faster screening for potential adaptive mechanisms by the use of high-throughput technologies [139]. In particular, during the last 10 years, apart from the gel-based techniques (e.g., 2D-PAGE and 2D-DIGE), gel-free techniques (e.g., stable-isotope labeling or using label-free methods) have been dominating the field of MS-based quantitation in proteomics [140]. This enhances the ability of proteomics to explore disease mechanisms. As mentioned above, proteomics analysis has been used to investigate the toxicity mechanism of fluorine on sperm [89]. Proteomics analysis associated with F-toxicity has also been studied in other tissues including gastrocnemius muscle, kidney, liver, midgut, bone, cells, serum, and urine [18, 54, 62, 68, 141–147]. All of these studies are listed in **Table 2**. As shown in **Table 2**, proteomic techniques 2D-PAGE, LC-MS/MS (liquid chromatography-tandem mass spectrometry), and iTRAQ (isobaric tags for relative and absolute quantification) labeling coupled with LC-MS/MS analysis were employed in these studies. The proteins associated with fluoride exposure were found involved in oxidative stress, ER stress, cell proliferation and apoptosis, mitochondrial-metabolism, tricarboxylic acid (TCA) cycle, unfolded protein response, inflammatory response, etc. These pathways or biological processes have previously been linked to the pathophysiology of fluorosis. The results support the current views on the molecular mechanism of F-toxicity.

Interestingly, Khan et al. evaluated the effects of F on the liver proteome of mice susceptible (A/J) or resistant (129P3/J) to the effects of F. As compared with 129P3/J mice, most of the proteins with fold change upon treatment with lower F concentrate (15 ppm) were increased in the A/J mice, suggesting an attempt of the former to fight the deleterious effects of F. However, upon treatment with 50 ppm F, most proteins with fold change were decreased in the A/J mice, especially proteins related to oxidative stress and protein folding, which might be related to the higher susceptibility of the A/J animals to the deleterious effects of F [18]. These results add light to the mechanisms underlying genetic susceptibility to fluorosis [18].

It is worth mentioning that in our previous comparative proteomic analysis of fluoride treated rat bone [54], 13, 35, and 34 differentially expressed proteins were identified in low-, medium-, and high-dose NaF exposure group. The medium- and high-dose groups shared a more similar protein expression pattern. Most of these proteins belong to collagen proteins, matrix metalloproteinases, proteoglycans (PGs), proteolytic protein, osteoclast-related protein, and myosin proteins, involved in collagen metabolism, bone mass change, mineralization process, dysfunction of the motor cell, and affected osteoblasts and/or osteoclasts, finally, contributing to the pathophysiology of skeletal and chronic fluorosis [54].

Models/materials	Toxic effects	Method	Critical proteins	Pathways/biology processes	References
Calvarial osteoblasts	Cells were treated with NaF	2-DE MALDI-TOF MS	ATP synthase, Dihydropyrimidinase-like 2, Heat shock 70-kDa protein(HSP70), Nucleoside diphosphate kinase, Glutamate oxaloacetate transaminase, Phosphatidylethanolamine binding prote in Proteasome 26S ATPase, Nucleoside diphosphate kinase, Protein disulfide isomerase, Ras-GTPase-activating protein, Thioredoxin, Tubulin, beta	Cell proliferation Nucleotide metabolism Signal transduction Protein oxidative folding Hydrophobic ligands Cell motility	Xu et al. [141]
Urine from the fluoride-treated Wistar rat	Dental fluorosis	2D-PAGE MALDI-TOF-TOF MS/MS	Androgen regulated 20 kDa protein, Aflatoxin B1 aldehyde reductase, alpha-2-μ-globulin	Detoxification Hormone regulation	Kobayashi et al [142]
Kidney (A/J and 129P3/J mice)	Dental fluorosis Kidney impairment Genetic susceptibility	2D-PAGE LC-MS/MS	Twenty five (25), 30 and 32 differentially expressed proteins were successfully identified between different doses of NaF treatment groups and their respective controls.	Metabolic and cellular processes Response to stimuli Development Regulation of cellular processes	Carvalho et al. [143]
Gastrocnemius muscle Streptozotocin-induced diabetes exposed to fluorides	Alter glucose homeostasis and lead to insulin resistance	2D-PAGE LC-MS/MS	78 kDa glucose-regulated protein, Alpha-enolase, Beta-enolase, Gamma-enolase, Gelsolin, Glyceraldehyde-3-phosphate dehydrogenase, Glycerol-3-phosphate dehydrogenase [NAD(+)], Heat shock cognate 71 kDa protein, L-lactate dehydrogenase A chain, L-lactate dehydrogenase B chain, L-lactate dehydrogenase C chain, Malate dehydrogenase, Myosin-3, Myosin-6, Myosin-7, Myosin-8, Myosin-binding protein C_ slow-type, Myosin-binding protein H, Pyruvate kinase isozymes M1/M2	Muscle contraction Carbohydrate catabolic processes Generation of precursor Metabolites and energy NAD metabolic processes Gluconeogenesis	Leite et al. [144]
Femurs, tibiae, and lumbar vertebrae (129P3/J and A/J mice)	Bone architecture Mineral apposition rate Genetic susceptibility	LC-ESI-MS/MS	129P3/J vs A/J mice: Bone morphogenetic protein 1, Bone sialoprotein 2, Collagen alpha-1(I) chain, Collagen alpha-2(I) chain, Exportin-2, NADPH oxidase 4, Protocadherin beta 15, Secreted frizzled-related sequence protein 4 129P3/J vs F-treated 129P3/J mice: Aflatoxin B1 aldehyde reductase member 2, Carbonyl reductase [NADPH] 2, Catenin alpha-2, Chromodomain-helicase-DNA-binding protein	Osteogenesis Osteoclastogenesis	Kobayashi et al. [62]

Models/materials	Toxic effects	Method	Critical proteins	Pathways/biology processes	References
			4, Chromodomain-helicase-DNA-binding protein 7, NADPH oxidase 4, Phosphatidylinositol 3,4,5-trisphosphate 5-phosphatase 2, Protocadherin beta 9 A/J vs F-treated A/J mice: Eukaryotic translation initiation factor 2 alpha kinase 3, Exportin-2, Phosphatidylinositol 3,4,5-trisphosphate 5-phosphatase 1		
Kidney (Wistar rats)	Alteration of renal metabolism	2D-PAGE MALDI-TOF MS	Control vs 10 ppm F: Aldo–keto reductase, Adnylate kinase 3-like 1, Enoyl coenzyme A hydratase, Pyruvate carboxylase, Control vs 5 ppm F: Aldolase B, Endoplasmic reticulum protein 29	Detoxification Metabolism Housekeeping	Kobayashi et al [145]
Hippocampus from rats	Leaning ability Memory	2D-PAGE, MALDI-TOF-MS	5'-AMP-activated protein kinase, Aconitate hydratase, Actin, cytoplasmic2-like isoform 3, Actr2 protein, Beta-actin, Cytosolic aspartate, dehydrogenase [NADP+], and fructose-bisphosphate aldolase C, Dynamin, Fascin, Fructose-bisphosphate aldolase C-B, Gln synthetase, Glycogen phosphorylase, glycoprotein 1 precursor, Lysosome-associated membrane, MHC class I antigen, Mitogen-activated protein kinase 1, Mixture 1: alcohol, N-ethylmaleimide sensitive, Otub1 protein, PDZ and LIM domain prptein 3, Phosphatase 1E isoform 1, Pyruvate carboxylase, Serum albumin, Tropomyosin 1, Ulip2 protein, Voltage-dependent anion-selective channel protein 1, WD repeat-containing protein 1	Biosynthesis of amino acids Carbon metabolism Insulin signaling pathway Phagosome Oxytocin signaling pathway	Pan et al. [146]
Serum from Wistar rats treated by NaF	Dental fluorosis	iTRAQ labeling, NanoLC-MS/MS	A total of 37 differentially expressed proteins were identified in different doses of the NaF treatment group	Complement and coagulation cascade Inflammatory response Complement activation Defense response Wound response	Wei et al. [147]

DOI: http://dx.doi.org/10.5772/intechopen.84548

Models/materials	Toxic effects	Method	Critical proteins	Pathways/biology processes	References
Sperm samples from Kunming mice	Sperm damage	2D-PAGE, MALDI-TOF-MS	Adenylate kinase isoenzyme 1 isoform 2, Aldose reductase-related protein 1 Annexin A13, Annexin A4, Dihydrolipoyllysine-residue acetyltransferase component of pyruvate dehydrogenase complex, Gamma-actin, Inhibitor alpha1 Phosphoglycerate kinase 2, Proteasome (prosome, macropain) subunit, alpha type 3, Serotransferrin precursor, Triosephosphate isomerase	Sperm motility Maturation Capacitation and acrosome reaction Lipid peroxidation Detoxification Inflammation Stability of membrane structure	Sun et al. [89]
Bone samples (Sprague-Dawley rats)	Chronic fluorosis	iTRAQ labeling, NanoLC-MS/MS	Thirteen (13), 35, and 34 differentially expressed proteins were identified in low-, medium-, and high-dose NaF-treated group, respectively. These proteins belong to collagen proteins, matrix metalloproteinases (MPPs), proteoglycans(PGs), proteolytic protein, osteoclast related protein, and myosin proteins	Bone mass change Mineralization process Dysfunction of the motor cell Affected on osteoblasts and/or osteoclasts	Wei et al. [54]
Liver (A/J, 129P3/J mice)	Disturbances in soft tissues Genetic susceptibility	Nano-LC-ESI-MS/MS	Eighty one (81) differentially expressed proteins were identified in the liver of A/J and 129P3/J mice treated with 15 ppm F. One hundred one (101) differentially expressed proteins were identified in the liver of A/J and 129P3/J mice treated with 50 ppm F	Carboxylic acid metabolic process Cellular amino acid metabolic Process Oxidative stress and protein folding might be related to the susceptibility to the deleterious effects of F	Khan et al. [18]
Plasma from Children	Children intelligence Gene polymorphism	2-DE, MALDI-TOF/TOF-MS	Alpha-1-B glycoprotein, Apolipoprotein E precursor, Complement C1s subcomponent precursor, Hemopexin, Immunoglobulin light chain variable region	Cell immunity Metabolism	Zhang et al. [68]

Table 2.
Fluorosis-related proteomics studies reported in the literatures.

Metabolomics can capture low-molecular weight metabolites that are the closest to the phenotype, which is believed to be one of the most powerful techniques to study the metabolic alteration associated with the treatment of environmental toxicants. However, the study on metabolic profile response to fluoride exposure is limited. A recent study carried out a metabonomics study on NaF treated human oral squamous cell carcinoma cells. The results showed that inhibition of the enolase reaction in glycolysis pathway was observed in the early stages of fluoride treatment. In the later stages, gradual increases in the AMP/ATP ratio (a putative marker of apoptosis) and oxidized products (e.g., GSSH, and methionine sulfoxide), and marginal changes in polyamine levels (putative marker of necrosis), were observed [148]. It suggested that the inhibition of enolase reaction and TCA cycle progression at early stage is specific to NaF, whereas the increase of ATP utilization at later stage may be common to apoptotis-inducing agents, but not to necrosis-inducing agents [148].

8. Treatment and prevention of chronic fluorosis

So far, there is no specific treatment for fluorosis. Efforts are being made to reduce the severity of the disease and improve quality of life of affected patients [149]. Medical treatment being used is mainly supplementation of vitamin (Vit) C, D, and E, calcium, antioxidants and treatment of malnutrition [150]. In recent years, some traditional Chinese medicines (TCM) have been developed to treat fluoride-induced bone lesions in China [49]. Treatment options for dental and skeletal fluorosis vary according to the severity of the disease [149]. Methods for treating dental fluorosis include micro/macro abrasion, bleaching, composite restorations, veneers, and full crowns [151]. Treatment of skeletal fluorosis may include surgical processes while treatment of deformity includes use of physiotherapy, corrective plasters, and orthoses (appropriate appliances) [149].

Clearly, chronic fluorosis is mainly caused by excess intake of fluoride through drinking water, food products, air, and industrial pollutants over a long period. Therefore, avoiding excessive intake of the fluoride is essential for the prevention of this disease. Notably, to keep fluoride intake within safe limits, one needs to consider the total daily intake, including fluoride intake from water, food, air, fluoride-rich dental products and drugs. The recommended level for daily fluoride intake is 0.05–0.07 mg F/Kg/day [152, 153]. In China, for people aged 8–16 (including 16-year-olds), the recommended values of the total daily fluoride intake per person is $\leq$2.4 mg; for those 16 years old, the total daily fluoride intake per person is $\leq$3.5 mg [154]. In water-borne fluorosis endemic areas, alternative water resources with low fluoride levels or defluorinated water can be used. Coal-burning endemic fluorosis areas need to change the way coal is burned and food is dried. Likewise, it is beneficial for a daily intake of foods, vegetables, and fruits rich in vitamin C, D, and E, calcium, and antioxidants for the prevention of chronic fluorosis in endemic regions [49, 82, 150]. Moreover, health education is a very important aspect for disease management. Knowledge regarding the harmful effects of fluoride and the causes of fluorosis can help people, especially the affected population, pay more attention to their living habits [49, 149]. Furthermore, the identification of candidate genes that affect risk factors is necessary to develop more effective measures to prevent and treat fluorosis [27].

9. Conclusions

The contents of this chapter are reviewed in **Figure 1**. Excess intake of fluoride can cause chronic fluorosis, leading to dental fluorosis and skeletal fluorosis and damage

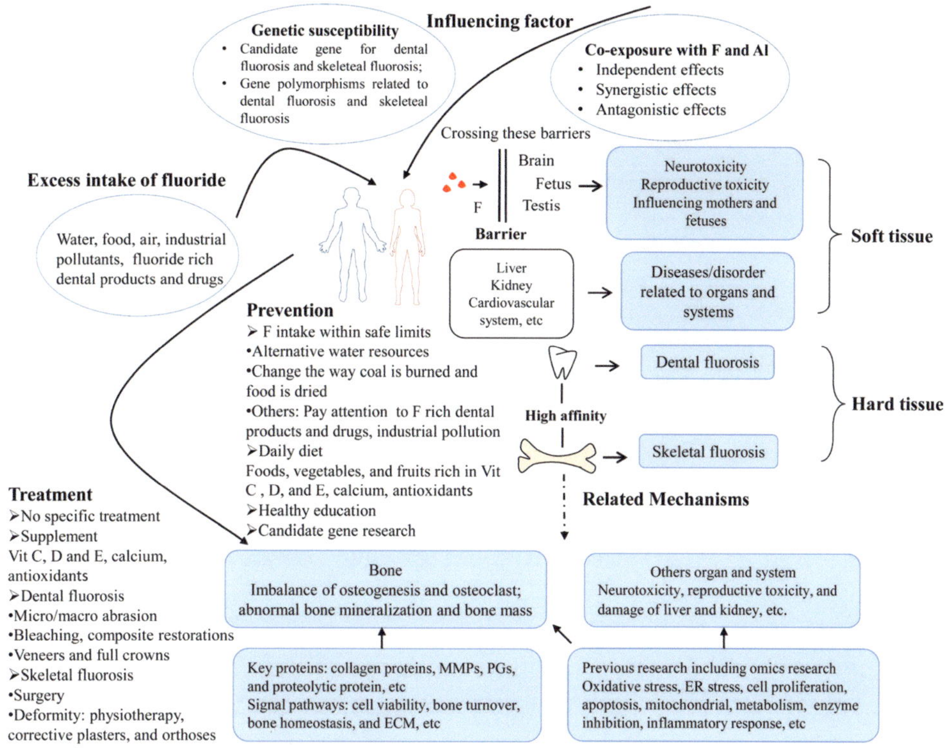

Figure 1.
An overview of the occurrence, influencing factors, pathogenesis, treatment, and prevention of fluorosis.

to nervous system, reproductive system, cardiovascular system, liver, and kidney. The possible mechanisms involved different key proteins and signal transduction pathways associated with the pathogenesis of fluorosis have been proposed. Some high-throughput methods such as proteomics, metabolomics, and transcriptomics have been used in the study of the mechanism underlying development of fluorosis. Genetic factors play a critical role in the pathogenesis of chronic fluorosis. Combined exposure to fluoride with other element such as arsenic or aluminum may result in more complicated adverse health effects than exposure to fluoride or these elements alone. Further research is needed to reveal the interaction between fluorides with these elements with regard to their toxic effects. Clearly, the mechanisms of chronic fluorosis still need further research. Prevention of chronic fluorosis is important and it can be prevented by keeping fluoride intake within safe limits. It is important to consider total exposure (i.e., exposure through air, food, and water) when evaluating adverse health effects of fluoride.

Acknowledgements

The authors would like to acknowledge the National Natural Science Foundation of China (Grant Nos. 31870825, 81560515) and Shenzhen Bureau of Science, Technology and Information (Nos. JCYJ20170412110026229, JCYJ20150529164656093) for funds to support this work.

Conflict of interest

The authors declare that they have no competing interests.

Author details

Liming Shen[1*], Chengyun Feng[2], Sijian Xia[1], Yan Wei[3], Hua Zhang[3], Danqing Zhao[4], Fang Yao[1], Xukun Liu[1], Yuxi Zhao[1] and Huajie Zhang[1]

1 College of Life Science and Oceanography, Shenzhen University, Shenzhen, PR China

2 Maternal and Child Health Hospital of Baoan, Shenzhen, PR China

3 School of Public Health, Guizhou Medical University, Guiyang, PR China

4 Department of Obstetrics and Gynecology, Affiliated Hospital of Guizhou Medical University, Guiyang, PR China

*Address all correspondence to: slm@szu.edu.cn

References

[1] Kanduti D, Sterbenk P, Artnik B. Fluoride: A review of use and effects on health. Mater Socio-Medica. 2016;**28**:133-137. DOI: 10.5455/msm.2016.28.133-137

[2] Clark MB, Slayton RL. Section on oral health. Fluoride use in caries prevention in the primary care setting. Pediatrics. 2014;**134**:626-633. DOI: 10.1542/peds.2014-1699

[3] Ozsvath DL. Fluoride and environmental health: A review. Reviews in Environmental Science and Biotechnology. 2009;**8**:59-79. DOI: 10.1007/s11157-008-9136-9

[4] Ullah R, Zafar MS, Shahani N. Potential fluoride toxicity from oral medicaments: A review. Iranian Journal of Basic Medical Sciences. 2017;**20**: 841-848. DOI: 10.22038/IJBMS. 2017.9104

[5] Everett ET. Fluoride's effects on the formation of teeth and bones, and the influence of genetics. Journal of Dental Research. 2011;**90**:552-560. DOI: 10.1177/0022034510384626

[6] Wu C, Gu X, Ge Y, Zhang J, Wang J. Effects of high fluoride and arsenic on brain biochemical indexes and learning-memory in rats. Fluoride. 2006;**39**:274-279

[7] Liu J, Gao Q, Wu CX, Guan ZZ. Alterations of nAChRs and ERK1/2 in the brains of rats with chronic fluorosis and their connections with the decreased capacity of learning and memory. Toxicology Letters. 2010;**192**:324-329. DOI: 10.1016/j.toxlet.2009.11.002

[8] Pereira M, Dombrowski PA, Losso EM, Chioca LR, Da Cunha C, Andreatini R. Memory impairment induced by sodium fluoride is associated with changes in brain monoamine levels. Neurotoxicity Research. 2011;**19**:55-62. DOI: 10.1007/s12640-009-9139-5

[9] Niu R, Xue X, Zhao Y, Sun Z, Yan X, Li X, et al. Effects of fluoride on microtubule ultrastructure and expression of Tubα1a and Tubβ2a in mouse hippocampus. Chemosphere. 2015;**139**:422-427. DOI: 10.1016/j.chemosphere.2015.07.011

[10] Wei N, Dong YT, Deng J, Wang Y, Qi XL, Yu WF, et al. Changed expressions of N-methyl-d-aspartate receptors in the brains of rats and primary neurons exposed to high level of fluoride. Journal of Trace Elements in Medicine and Biology: Organ of the Society for Minerals and Trace Elements (GMS). 2018;**45**:31-40. DOI: 10.1016/j.jtemb.2017.09.020

[11] Dong YT, Wang Y, Wei N, Zhang QF, Guan ZZ. Deficit in learning and memory of rats with chronic fluorosis correlates with the decreased expressions of M1 and M3 muscarinic acetylcholine receptors. Archives of Toxicology. 2015;**89**:1981-1991. DOI: 10.1007/s00204-014-1408-2

[12] Pain G. Mechanisms of fluoride neurotoxicity a quick guide to the literature. 2017. Available from: https://www.researchgate.net/publication/312057754_Mechanisms_of_Fluoride_Neurotoxicity_A_quick_guide_to_the_literature

[13] Chandrajith R, Dissanayake CB, Ariyarathna T, Herath HMJMK, Padmasiri JP. Dose-dependent Na and Ca in fluoride-rich drinking water--another major cause of chronic renal failure in tropical arid regions. Science of the Total Environment. 2011;**409**:671-675. DOI: 10.1016/j.scitotenv.2010.10.046

[14] Chattopadhyay A, Podder S, Agarwal S, Bhattacharya S.

Fluoride-induced histopathology and synthesis of stress protein in liver and kidney of mice. Archives of Toxicology. 2011;**85**:327-335. DOI: 10.1007/s00204-010-0588-7

[15] Pain G. Fluoride is a developmental nephrotoxin—Coming to a Kidney near you. 2017. Available from: https://www.researchgate.net/publication/313025968_Fluoride_is_a_developmental_Nephrotoxin_-_coming_to_a_Kidney_near_you. DOI: 10.13140/RG.2.2.10999.62884

[16] Iano FG, Ferreira MC, Quaggio GB, Fernandes MS, Oliveira RC, Ximenes VF, et al. Effects of chronic fluoride intake on the antioxidant systems of the liver and kidney in rats. Journal of Fluorine Chemistry. 2014;**168**:212-217. DOI: 10.1016/j.jfluchem.2014.09.029

[17] Zhao Y, Li Y, Wang J, Manthari RK, Wang J. Fluoride induces apoptosis and autophagy through the IL-17 signaling pathway in mice hepatocytes. Archives of Toxicology. 2018;**92**:3277-3289. DOI: 10.1007/s00204-018-2305-x

[18] Khan ZN, Sabino IT, de Souza Melo CG, Martini T, da Silva Pereira HAB, Buzalaf MAR. Liver proteome of mice with distinct genetic susceptibilities to fluorosis treated with different concentrations of F in the drinking water. Biological Trace Element Research. 2019;**187**:107-119. DOI: 10.1007/s12011-018-1344-8

[19] Fan B, Yu Y, Zhang Y. PI3K-Akt1 expression and its significance in liver tissues with chronic fluorosis. International Journal of Clinical and Experimental Pathology. 2015;**8**:1226-1236

[20] Amini H, Taghavi Shahri SM, Amini M, Ramezani Mehrian M, Mokhayeri Y, Yunesian M. Drinking water fluoride and blood pressure? An environmental study. Biological Trace Element Research. 2011;**144**:157-163. DOI: 10.1007/s12011-011-9054-5

[21] Sun L, Gao Y, Liu H, Zhang W, Ding Y, Li B, et al. An assessment of the relationship between excess fluoride intake from drinking water and essential hypertension in adults residing in fluoride endemic areas. Science of the Total Environment. 2013;**443**:864-869. DOI: 10.1016/j.scitotenv.20-12.11.021

[22] Liu H, Gao Y, Sun L, Li M, Li B, Sun D. Assessment of relationship on excess fluoride intake from drinking water and carotid atherosclerosis development in adults in fluoride endemic areas, China. International Journal of Hygiene and Environmental Health. 2014;**217**:413-420. DOI: 10.1016/j.ijheh.2013.08.001

[23] Sun L, Gao Y, Zhang W, Liu H, Sun D. Effect of high fluoride and high fat on serum lipid levels and oxidative stress in rabbits. Environmental Toxicology and Pharmacology. 2014;**38**:1000-1006. DOI: 10.1016/j.et-ap.2014.10.010

[24] Huo M, Han H, Sun Z, Lu Z, Yao X, Wang S, et al. Role of IL-17 pathways in immune privilege: A RNA deep sequencing analysis of the mice testis exposure to fluoride. Scientific Reports. 2016;**6**:32173. DOI: 10.1038/srep32173

[25] He X, Sun Z, Manthari RK, Wu P, Wang J. Fluoride altered rat's blood testis barrier by affecting the F-actin via IL-1α. Chemosphere. 2018;**211**:826-833. DOI: 10.1016/j.chemosphere.20-18.08.009

[26] Han H, Sun Z, Luo G, Wang C, Wei R, Wang J. Fluoride exposure changed the structure and the expressions of reproductive related genes in the hypothalamus-pituitary-testicular axis of male mice. Chemosphere. 2015;**135**:297-303. DOI: 10.1016/j.chemosphere.2015.04.012

[27] Pramanik S, Saha D. The genetic influence in fluorosis. Environmental

Toxicology and Pharmacology. 2017;**56**:157-162. DOI: 10.1016/j.etap.2017.09.008

[28] Xu Y, Huang H, Zeng Q, Yu C, Yao M, Hong F, et al. The effect of elemental content on the risk of dental fluorosis and the exposure of the environment and population to fluoride produced by coal-burning. Environmental Toxicology and Pharmacology. 2017;**56**:329-339. DOI: 10.1016/j.etap.2017.10.011

[29] Qin X, Wang S, Yu M, Zhang L, Li X, Zuo Z, et al. Child skeletal fluorosis from indoor burning of coal in southwestern China. Journal of Environmental and Public Health. 2009;**2009**:969764. DOI: 10.1155/2009/9-69764

[30] Chen J, Liu G, Kang Y, Wu B, Sun R, Zhou C, et al. Coal utilization in China: Environmental impacts and human health. Environmental Geochemistry and Health. 2014;**36**:735-753. DOI: 10.1007/s10653-013-9592-1

[31] Wang C, Gao Y, Wang W, Zhao L, Zhang W, Han H, et al. A national cross-sectional study on effects of fluoride-safe water supply on the prevalence of fluorosis in China. BMJ Open. 2012;**2**:e001564. DOI: 10.1136/bmjopen-2012-001564

[32] Fan Z, Gao Y, Wang W, Gong H, Guo M, Zhao S, et al. Prevalence of brick tea-type fluorosis in the Tibet autonomous region. Journal of Epidemiology. 2016;**26**:57-63. DOI: 10.2188/jea.JE20150037

[33] Linhares D, Camarinho R, Garcia PV, Rodrigues ADS. Mus musculus bone fluoride concentration as a useful biomarker for risk assessment of skeletal fluorosis in volcanic areas. Chemosphere. 2018;**205**:540-544. DOI: 10.1016/j.chemosphere.2018.04.144

[34] Ferreira AF, Garcia PV, Camarinho R, Rodrigues A dos S. Volcanogenic pollution and testicular damage in wild mice. Chemosphere. 2015;**132**:135-141. DOI: 10.1016/j.chemosphere.20-15.03.0-17

[35] Agalakova N, Gusev G. Molecular mechanisms of cytotoxicity and apoptosis induced by inorganic fluoride. ISRN Cell Biology. 2012;**2012**. Article ID 403835. DOI: 10.5402/2012/403835

[36] Dhar V, Bhatnagar M. Physiology and toxicity of fluoride. Indian Journal of Dental Research. 2009;**20**:350-355. DOI: 10.4103/0970-9290.57379

[37] Perumal E, Paul V, Govindarajan V, Panneerselvam L. A brief review on experimental fluorosis. Toxicology Letters. 2013;**223**:236-251. DOI: 10.1016/j.toxlet.2013.09.005

[38] Barbier O, Arreola-Mendoza L, Del Razo LM. Molecular mechanisms of fluoride toxicity. Chemico-Biological Interactions. 2010;**188**:319-333. DOI: 10.1016/j.cbi.2010.07.011

[39] Zuo H, Chen L, Kong M, Qiu L, Lü P, Wu P, et al. Toxic effects of fluoride on organisms. Life Sciences. 2018;**198**:18-24. DOI: 10.1016/j.lfs.2018.02.001

[40] Krishnamachari KA. Skeletal fluorosis in humans: A review of recent progress in the understanding of the disease. Progress in Food & Nutrition Science. 1986;**10**:279-314

[41] Topuz O, Akkaya N, Ardic F, Sarsan A, Cubukcu D, Gokgoz A. Bone resorption marker and ultrasound measurements in adults residing in an endemic fluorosis area of Turkey. Fluoride. 2006;**39**:138-144

[42] Mousny M, Omelon S, Wise L, Everett ET, Dumitriu M, Holmyard DP, et al. Fluoride effects on bone formation and mineralization are influenced by genetics. Bone. 2008;**43**:1067-1074. DOI: 10.1016/j.bone.2008.07.248

[43] Song Y, Tan H, Liu K, Zhang Y, Liu Y, Lu C, et al. Effect of fluoride exposure on bone metabolism indicators ALP, BALP, and BGP. Environmental Health and Preventive Medicine. 2011;**16**:158-163. DOI: 10.1007/s12199-010-0181-y

[44] Jianping Z, Benli Y. The application of bone metabolism biochemical markers in the study of skeletal fluorosis. Chinese Journal of Endemiology. 2003;**22**:186-188

[45] Debiński A, Nowicka G. Effect of sodium fluoride on ectopic induction of bone tissue. Annales Academiae Medicae Stetinensis. 2004;**50**(Suppl 1): 23-27

[46] Ma L, Lin ZQ, Liu KT. The effects of sodium fluoride on the resorptive activity of isolated osteoclasts. Journal of Xinjiang Medical University. 2004;**27**:334-336 (In Chinese)

[47] Okuda A, Kanehisa J, Heersche JN. The effects of sodium fluoride on the resorptive activity of isolated osteoclasts. Journal of Bone and Mineral Research. 1990;**5**(Suppl 1):S115-S120. DOI: 10.1002/jbmr.5650051381

[48] Junrui P, Bingyun L, Yanhui G, Xu J, Darko GM, Dianjun S. Relationship between fluoride exposure and osteoclast markers during RANKL-induced osteoclast differentiation. Environmental Toxicology and Pharmacology. 2016;**46**:241-245. DOI: 10.1016/j.etap.2016.08.001

[49] Chen Y, Yan W, Hui X. Treatment and prevention of skeletal fluorosis. Biomedical and Environmental Sciences. 2017;**30**:147-149. DOI: 10.3967/bes2017.020

[50] Lai LP, Lotinun S, Bouxsein ML, Baron R, McMahon AP. Stk11 (Lkb1) deletion in the osteoblast lineage leads to high bone turnover, increased trabecular bone density and cortical porosity. Bone. 2014;**69**:98-108. DOI: 10.1016/j.bone.2014.09.010

[51] Ito M, Nakagawa H, Okada T, Miyazaki S, Matsuo S. ER-stress caused by accumulated intracistanal granules activates autophagy through a different signal pathway from unfolded protein response in exocrine pancreas cells of rats exposed to fluoride. Archives of Toxicology. 2009;**83**:151-159. DOI: 10.1007/s00204-008-0341-7

[52] Daiwile AP, Sivanesan S, Tarale P, Naoghare PK, Bafana A, Parmar D, et al. Role of fluoride induced histone trimethylation in development of skeletal fluorosis. Environmental Toxicology and Pharmacology. 2018;**57**:159-165. DOI: 10.1016/j.etap.2017.12.015

[53] Iwatsuki M, Matsuoka M. Fluoride-induced c-Fos expression in MC3T3-E1 osteoblastic cells. Toxicology Mechanisms and Methods. 2016;**26**:132-138. DOI: 10.3109/15376516.2015.1129570

[54] Wei Y, Zeng B, Zhang H, Chen C, Wu Y, Wang N, et al. Comparative proteomic analysis of fluoride treated rat bone provides new insights into the molecular mechanisms of fluoride toxicity. Toxicology Letters. 2018;**291**:39-50. DOI: 10.1016/j.toxlet.2018.04.006

[55] Miao Q, Xu M, Liu B, You B. In vivo and in vitro study on the effect of excessive fluoride on type I collagen of rats. Wei Sheng Yan Jiu. 2002;**31**:145-147

[56] Yan X, Feng C, Chen Q, Li W, Wang H, Lv L, et al. Effects of sodium fluoride treatment in vitro on cell proliferation, apoptosis and caspase-3 and caspase-9 mRNA expression by neonatal rat osteoblasts. Archives of Toxicology. 2009;**83**:451-458. DOI: 10.1007/s00204-008-0365-z

[57] Yan X, Yan X, Morrison A, Han T, Chen Q, Li J, et al. Fluoride

induces apoptosis and alters collagen I expression in rat osteoblasts. Toxicology Letters. 2011;**200**:133-138. DOI: 10.1016/j.toxlet.20-10.11.005

[58] Yan X, Hao X, Nie Q, Feng C, Wang H, Sun Z, et al. Effects of fluoride on the ultrastructure and expression of type I collagen in rat hard tissue. Chemosphere. 2015;**128**:36-41. DOI: 10.1016/j.chemosphere.2014.12.090

[59] Zhong DB, Sun LT, Wu PF, Su JL, Han B. Effect of fluoride on proliferation, differentiation, and apoptosis of cultured caprine (goat) osteoblasts. Fluoride. 2005;**38**:230-231

[60] Li W, Yang L, Ren Y, Yan X, Wang J. Quantification of rib COL1A2 gene expression in healthy and fluorosed Inner Mongolia cashmere goats. Fluoride. 2007;**40**:13-18

[61] Nair M, Belak ZR, Ovsenek N. Effects of fluoride on expression of bone-specific genes in developing *Xenopus laevis* larvae. Biochemistry and Cell Biology. 2011;**89**:377-386. DOI: 10.1139/o11-034

[62] Kobayashi CAN, Leite AL, Peres-Buzalaf C, Carvalho JG, Whitford GM, Everett ET, et al. Bone response to fluoride exposure is influenced by genetics. PLoS One. 2014;**9**:e114343. DOI: 10.1371/journal.pone.0114343

[63] Waddington RJ, Langley MS. Structural analysis of proteoglycans synthesized by mineralizing bone cells in vitro in the presence of fluoride. Matrix Biology: Journal of the International Society for Matrix Biology. 1998;**17**:255-268

[64] Slompo C, Buzalaf CP, Damante CA, Martins GM, Hannas AR, Buzalaf MAR, et al. Fluoride modulates preosteoblasts viability and matrix metalloproteinases-2 and -9 activities. Brazilian Dental Journal. 2012;**23**:629-634

[65] Waddington RJ, Langley MS. Altered expression of matrix metalloproteinases within mineralizing bone cells in vitro in the presence of fluoride. Connective Tissue Research. 2003;**44**:88-95

[66] Dec K, Łukomska A, Maciejewska D, Jakubczyk K, Baranowska-Bosiacka I, Chlubek D, et al. The influence of fluorine on the disturbances of homeostasis in the central nervous system. Biological Trace Element Research. 2017;**177**:224-234. DOI: 10.1007/s12011-016-0871-4

[67] Choi AL, Sun G, Zhang Y, Grandjean P. Developmental fluoride neurotoxicity: A systematic review and meta-analysis. Environmental Health Perspectives. 2012;**120**:1362-1368. DOI: 10.1289/ehp.1104912

[68] Zhang S, Zhang X, Liu H, Qu W, Guan Z, Zeng Q, et al. Modifying effect of COMT gene polymorphism and a predictive role for proteomics analysis in children's intelligence in endemic fluorosis area in Tianjin, China. Toxicological Sciences. 2015;**144**: 238-245. DOI: 10.1093/toxsci/kfu311

[69] Choi AL, Zhang Y, Sun G, Bellinger DC, Wang K, Yang XJ, et al. Association of lifetime exposure to fluoride and cognitive functions in Chinese children: A pilot study. Neurotoxicology and Teratology. 2015;**47**:96-101. DOI: 10.1016/j.ntt.2014.11.001

[70] Whitford GM, Whitford JL, Hobbs SH. Appetitive-based learning in rats: Lack of effect of chronic exposure to fluoride. Neurotoxicology and Teratology. 2009;**31**:210-215. DOI: 10.1016/j.ntt.20-09.02.003

[71] Gao Q, Liu YJ, Guan ZZ. Decreased learning and memory ability in rats with fluorosis: Increased oxidative stress and reduced cholinesterase activity in the brain. Fluoride. 2009;**42**:277-285

[72] Liu F, Ma J, Zhang H, Liu P, Liu YP, Xing B, et al. Fluoride exposure during development affects both cognition and emotion in mice. Physiology & Behavior. 2014;**124**:1-7. DOI: 10.1016/j.physbeh.20-13.10.027

[73] Zhang Z, Xu X, Shen X, Xu X. Effect of fluoride exposure on synaptic structure of brain areas related to learning-memory in mice. Wei Sheng Yan Jiu. 1999;**28**:210-212 (In Chinese)

[74] Zhang KL, Lou DD, Guan ZZ. Activation of the AGE/RAGE system in the brains of rats and in SH-SY5Y cells exposed to high level of fluoride might connect to oxidative stress. Neurotoxicology and Teratology. 2015;**48**:49-55. DOI: 10.1016/j.ntt.2015.01.007

[75] Yan N, Liu Y, Liu S, Cao S, Wang F, Wang Z, et al. Fluoride-induced neuron apoptosis and expressions of inflammatory factors by activating microglia in rat brain. Molecular Neurobiology. 2016;**53**:4449-4460. DOI: 10.1007/s12035-015-9380-2

[76] Scheff SW, Price DA, Hicks RR, Baldwin SA, Robinson S, Brackney C. Synaptogenesis in the hippocampal CA1 field following traumatic brain injury. Journal of Neurotrauma. 2005;**22**:719-732. DOI: 10.1089/neu.2005.22.719

[77] Saxena R, Meena B, Chouhan VS, Bhatnagar M, et al. Biochemical changes in brain and other tissues of young adult female mice from fluoride in their drinking water. In: Fluoride (Res Rep November 1, 2). Vol. 39. 2006. pp. 280-284

[78] Chirumari K, Reddy PK. Dose-dependent effects of fluoride on neurochemical milieu in the hippocampus and neocortex of rat brain. Fluoride. 2007;**40**:101-110

[79] Gui CZ, Ran LY, Li JP, Guan ZZ. Changes of learning and memory ability and brain nicotinic receptors of rat offspring with coal burning fluorosis. Neurotoxicology and Teratology. 2010;**32**:536-541. DOI: 10.1016/j.ntt.2010.03.010

[80] Bartos M, Gumilar F, Bras C, Gallegos CE, Giannuzzi L, Cancela LM, et al. Neurobehavioural effects of exposure to fluoride in the earliest stages of rat development. Physiology & Behavior. 2015;**147**:205-212. DOI: 10.1016/j.physbeh.2015.04.044

[81] Schulz JA, Lamb AR. The effect of fluorine as sodium fluoride on the growth and reproduction of albino rats. Science. 1925;**61**:93-94. DOI: 10.1126/science.61.1569.93

[82] Tian Y, Xiao Y, Wang B, Sun C, Tang K, Sun F. Vitamin E and lycopene reduce coal burning fluorosis-induced spermatogenic cell apoptosis via oxidative stress-mediated JNK and ERK signaling pathways. Bioscience Reports. 2018;**38**:BSR20171003. DOI: 10.1042/BSR20171003

[83] Pati P, Bhunya S. Genotoxic effect of an environmental-pollutant, sodium-fluoride, in mammalian in vivo test system. Caryologia. 1987;**40**:79-87. DOI: 10.1080/00087114.1987.10797811

[84] Zhang J, Li Z, Qie M, Zheng R, Shetty J, Wang J. Sodium fluoride and sulfur dioxide affected male reproduction by disturbing blood-testis barrier in mice. Food and Chemical Toxicology. 2016;**94**:103-111. DOI: 10.1016/j.fct.2016.05.017

[85] Inkielewicz I, Krechniak J. Fluoride effects on glutathione peroxidase and lipid peroxidation in rats. Fluoride. 2004;**37**:7-12

[86] Dhurvey V, Patil V, Thakare M. Effect of sodium fluoride on the structure and function of the thyroid and ovary in albino rats (rattus Norvegicus). Fluoride. 2017;**50**:235-246

[87] Gofa A, Davidson RM. NaF potentiates a K (+)-selective ion channel in G292 osteoblastic cells. The Journal of Membrane Biology. 1996;**149**:211-219

[88] Wei R, Luo G, Sun Z, Wang S, Wang J. Chronic fluoride exposure-induced testicular toxicity is associated with inflammatory response in mice. Chemosphere. 2016;**153**:419-425. DOI: 10.1016/j.chemosphere.2016.03.045

[89] Sun Z, Wei R, Luo G, Niu R, Wang J. Proteomic identification of sperm from mice exposed to sodium fluoride. Chemosphere. 2018;**207**:676-681. DOI: 10.1016/j.chemosphere.2018.05.153

[90] Sun Z, Zhang W, Li S, Xue X, Niu R, Shi L, et al. Altered miRNAs expression profiling in sperm of mice induced by fluoride. Chemosphere. 2016;**155**:109-114. DOI: 10.1016/j.chemosphere.2016.04.053

[91] Buzalaf CP, Leite ADL, Buzalaf MAR. Fluoride Metabolism. In: Preedy VR, editor. Fluorine: Chemistry, Analysis, Function and Effects. Royal Society of Chemistry; 2015. p. 54-74. DOI: 10.10-39/9781782628507-00054

[92] Niu R, Han H, Sun Z, Zhang Y, Yin W, Wang J, et al. Effects of fluoride exposure on the antioxidative status in the kidneys. Fluoride. 2016;**49**:5-12

[93] Ma Y, Ma Z, Yin S, Yan X, Wang J. Arsenic and fluoride induce apoptosis, inflammation and oxidative stress in cultured human umbilical vein endothelial cells. Chemosphere. 2017;**167**:454-461. DOI: 10.1016/j.chemosphere.2016.10.025

[94] Dequeker J, Declerck K. Fluor in the treatment of osteoporosis. An overview of thirty years clinical research. Schweizerische Medizinische Wochenschrift. 1993;**123**:2228-2234

[95] Yoder KM, Mabelya L, Robison VA, Dunipace AJ, Brizendine EJ, Stookey GK. Severe dental fluorosis in a Tanzanian population consuming water with negligible fluoride concentration. Community Dentistry and Oral Epidemiology. 1998;**26**:382-393

[96] Choubisa SL, Choubisa L, Choubisa DK. Endemic fluorosis in Rajasthan. Indian Journal of Environmental Health. 2001;**43**:177-189

[97] Everett ET, McHenry MAK, Reynolds N, Eggertsson H, Sullivan J, Kantmann C, et al. Dental fluorosis: Variability among different inbred mouse strains. Journal of Dental Research. 2002;**81**:794-798. DOI: 10.1177/0810794

[98] Zeng BB, Zhang YF, Xia M, et al. Comparison of dental fluorosis susceptibility between SD rats and Wistar rats. Journal of Environment & Health. 2015;**32**:867-871 (In Chinese)

[99] Huang H, Ba Y, Cui L, Cheng X, Zhu J, Zhang Y, et al. COL1A2 gene polymorphisms (Pvu II and Rsa I), serum calciotropic hormone levels, and dental fluorosis. Community Dentistry and Oral Epidemiology. 2008;**36**:517-522. DOI: 10.1111/j.1600-0528.2007.00424

[100] Jarquín Yñezá L, Alegría-Torres JA, Castillo CG, de Jesús Mejía-Saavedra J. Dental fluorosis and a polymorphism in the COL1A2 gene in Mexican children. Archives of Oral Biology. 2018;**96**:21-25. DOI: 10.1016/j.archoralbio.2018.08.010

[101] Ba Y, Zhang H, Wang G, Wen S, Yang Y, Zhu J, et al. Association of dental fluorosis with polymorphisms of estrogen receptor gene in Chinese children. Biological Trace Element Research. 2011;**143**:87-96. DOI: 10.1007/s12011-010-8848-1

[102] Küchler EC, Dea Bruzamolin C, Ayumi Omori M, Costa MC, Antunes LS, Pecharki GD, et al. Polymorphisms

in nonamelogenin enamel matrix genes are associated with dental fluorosis. Caries Research. 2018;**52**:1-6. DOI: 10.1159/000479826

[103] Küchler EC, Tannure PN, de ODSB, Charone S, Nelson-Filho P, da Silva RAB, et al. Polymorphisms in genes involved in enamel development are associated with dental fluorosis. Archives of Oral Biology. 2017;**76**:66-69. DOI: 10.1016/j.archoralbio.2017.01.009

[104] Yang Y, Zhao Q, Liu Y, Liu X, Chu Y, Yan H, et al. FRZB1 rs2242070 polymorphisms is associated with brick tea type skeletal fluorosis in Kazakhs, but not in Tibetans, China. Archives of Toxicology. 2018;**92**:2217-2225. DOI: 10.1007/s00204-018-2217-9

[105] Escobar-García D, Mejía-Saavedra J, Jarquín-Yáñez L, Molina-Frechero N, Pozos-Guillén A. Collagenase 1A2 (COL1A2) gene A/C polymorphism in relation to severity of dental fluorosis. Community Dentistry and Oral Epidemiology. 2016;**44**:162-168. DOI: 10.1111/cdoe.12201

[106] Wu J, Wang W, Liu Y, Sun J, Ye Y, Li B, et al. Modifying role of GSTP1 polymorphism on the association between tea fluoride exposure and the brick-tea type fluorosis. PLoS One. 2015;**10**:e0128280. DOI: 10.1371/journal.pone.0128280

[107] Li BY, Yang YM, Liu Y, Sun J, Ye Y, Liu XN, et al. Prolactin rs1341239 T allele may have protective role against the brick tea type skeletal fluorosis. PLoS One. 2017;**12**:e0171011. DOI: 10.1371/journal.pone.0171011

[108] Pei J, Li B, Liu Y, Liu X, Li M, Chu Y, et al. Matrix metallopeptidase-2 gene rs2287074 polymorphism is associated with brick tea skeletal fluorosis in Tibetans and Kazaks, China. Scientific Reports. 2017;**7**:40086. DOI: 10.1038/srep40086

[109] Chouhan S, Flora SJS. Arsenic and fluoride: Two major ground water pollutants. Indian Journal of Experimental Biology. 2010;**48**:666-678

[110] Wang GQ, Huang YZ, Xiao BY, Qian XC, Yao H, Hu Y, et al. Toxicity from water containing arsenic and fluoride in Xinjiang. Fluoride. 1997;**30**:81-84

[111] González-Horta C, Ballinas-Casarrubias L, Sánchez-Ramírez B, Ishida MC, Barrera-Hernández A, Gutiérrez-Torres D, et al. A concurrent exposure to arsenic and fluoride from drinking water in Chihuahua, Mexico. International Journal of Environmental Research and Public Health. 2015;**12**:4587-4601. DOI: 10.3390/ijerph120504587

[112] Lin G, Gong S, Wei C, Chen J, Golka K, Shen J. Co-occurrence of arseniasis and fluorosis due to indoor combustion of high fluorine and arsenic content coal in a rural township in Northwest China: Epidemiological and toxicological aspects. Archives of Toxicology. 2012;**86**:839-847. DOI: 10.1007/s00204-011-0792-0

[113] An D, Ho GY, Hu XQ. Chronic arsenic-fluorine intoxication from burning coals with high arsenic and fluorine content. Chinese Journal of Preventive Medicine. 1994;**28**:312-313 (In Chinese)

[114] Zeng Q, Xu Y, Yu X, Yang J, Hong F, Zhang A. The combined effects of fluorine and arsenic on renal function in a Chinese population. Toxicology Research. 2014;**3**:359-366. DOI: 10.1039/C4TX00038B

[115] Zeng Q, Xu Y, Yu X, Yang J, Hong F, Zhang A. Arsenic may be involved in fluoride-induced bone toxicity through PTH/PKA/AP1 signaling pathway. Environmental Toxicology and Pharmacology. 2014;**37**:228-233. DOI: 10.1016/j.etap.2013.11.027

[116] Jia L, Jin TY. Combined effect of fluoride and arsenate on gene expression of osteoclast differentiation factor and osteoprotegerin. Biomedical and Environmental Sciences. 2006;**19**:375-379

[117] Ma Y, Niu R, Sun Z, Wang J, Luo G, Zhang J, et al. Inflammatory responses induced by fluoride and arsenic at toxic concentration in rabbit aorta. Archives of Toxicology. 2012;**86**:849-856

[118] Jiang S, Su J, Yao S, Zhang Y, Cao F, Wang F, et al. Fluoride and arsenic exposure impairs learning and memory and decreases mGluR5 expression in the hippocampus and cortex in rats. PLoS One. 2014;**9**:e96041. DOI: 10.1371/journal.pone.0096041

[119] Li FC. Report on disease caused by kaolin mixed with coal for roasting corn. Chinese Journal of Preventive Medicine. 1988;**22**:225-229 (In Chinese)

[120] Wong MH, Fung KF, Carr HP. Aluminium and fluoride contents of tea, with emphasis on brick tea and their health implications. Toxicology Letters. 2003;**137**:111-120

[121] Zhu Y, Xu F, Yan X, Miao L, Li H, Hu C, et al. The suppressive effects of aluminum chloride on the osteoblasts function. Environmental Toxicology and Pharmacology. 2016;**48**:125-129. DOI: 10.1016/j.etap.20-16.10.009

[122] Sun X, Liu J, Zhuang C, Yang X, Han Y, Shao B, et al. Aluminum trichloride induces bone impairment through TGF-β1/Smad signaling pathway. Toxicology. 2016;**371**:49-57. DOI: 10.1016/j.tox.2016.10.002

[123] Sun X, Wang H, Huang W, Yu H, Shen T, Song M, et al. Inhibition of bone formation in rats by aluminum exposure via Wnt/β-catenin pathway. Chemosphere. 2017;**176**:1-7. DOI: 10.1016/j.chemosphere.2017.02.086

[124] Yang X, Huo H, Xiu C, Song M, Han Y, Li Y, et al. Inhibition of osteoblast differentiation by aluminum trichloride exposure is associated with inhibition of BMP-2/Smad pathway component expression. Food and Chemical Toxicology. 2016;**97**:120-126. DOI: 10.1016/j.fct.2016.09.004

[125] Allain P, Gauchard F, Krari N. Enhancement of aluminum digestive absorption by fluoride in rats. Research Communications in Molecular Pathology and Pharmacology. 1996;**91**:225-231

[126] Zhang H, Wei Y, Xie C, et al. Experimental study on the relationship between fluorosis and different dosage proportion of fluorine to aluminum. Chinese Jouranl of Endemiology. 2001;**20**:426-428

[127] Lubkowska A, Chlubek D, Machoy-Mokrzyniska A. The effect of alternating administration of aluminum chloride and sodium fluoride in drinking water on the concentration of fluoride in serum and its content in bones of rats. Annales Academiae Medicae Stetinensis. 2006;**52** (Suppl 1):67-71

[128] Guo X, Cai R, Wu S, He Y, Sun G. Combined effect of fluoride and aluminum on the expression of Runx2 and Osterix mRNA in MC3T3-E1 cells. Wei Sheng Yan Jiu. 2011;**40**:164-166

[129] Lubkowska A, Zyluk B, Chlubeka D. Interactions between fluorine and aluminum. Fluoride. 2002;**35**:73-77

[130] Ghorbel I, Amara IB, Ktari N, Elwej A, Boudawara O, Boudawara T, et al. Aluminium and acrylamide disrupt cerebellum redox states, cholinergic function and membrane-bound ATPase in adult rats and their offspring. Biological Trace Element Research. 2016;**174**:335-346. DOI: 10.1007/s12011-016-0716-1

[131] Lubkowska A, Chlubeka D, Machoy-Mokrzynska A, et al. Distribution of fluoride in selected structures of the central nervous system in rats exposed to NaF and AlCl3 in drinking water. Trace Elements & Electrolytes. 2012;**29**:162-171

[132] Akinrinade ID, Memudu AE, Ogundele OM, Ajetunmobi OI. Interplay of glia activation and oxidative stress formation in fluoride and aluminium exposure. Pathophysiology. 2015;**22**:39-48. DOI: 10.1016/j.pathophys.2014.12.001

[133] Akinrinade ID, Memudu AE, Ogundele OM. Fluoride and aluminium disturb neuronal morphology, transport functions, cholinesterase, lysosomal and cell cycle activities. Pathophysiology, the Official Journal of the International Society for Pathophysiology. 2015;**22**:105-115. DOI: 10.1016/j.pathophys.20-15.03.001

[134] Ge QD, Xie C, Zhang H, Tan Y, Wan CW, Wang WJ, et al. Differential expression of miRNAs in the hippocampi of offspring rats exposed to fluorine combined with aluminum during the embryonic stage and into adulthood. Biological Trace Element Research. 2018:1-15. DOI: 10.1007/s12011-018-1445-4

[135] Ge QD, Tan Y, Luo Y, Wang WJ, Zhang H, Xie C. MiR-132, miR-204 and BDNF-TrkB signaling pathway may be involved in spatial learning and memory impairment of the offspring rats caused by fluorine and aluminum exposure during the embryonic stage and into adulthood. Environmental Toxicology and Pharmacology. 2018;**63**:60-68. DOI: 10.1016/j.etap.2018.08.011

[136] Chinoy NJ, Patel TN. Effects of sodium fluoride and aluminium chloride on ovary and uterus of mice and their reversal by some antidotes. Fluoride. 2001;**34**:9-20

[137] Dong C, Cao J, Cao C, Han Y, Wu S, Wang S, et al. Effects of fluoride and aluminum on expressions of StAR and P450scc of related steroidogenesis in Guinea pigs' testis. Chemosphere. 2016;**147**:345-351. DOI: 10.1016/j.chemosphere.2015.12.064

[138] Wasana HM, Perera GD, De Gunawardena PS, Bandara J. The impact of aluminum, fluoride, and aluminum–fluoride complexes in drinking water on chronic kidney disease. Environmental Science and Pollution Research. 2015;**22**:11001-11009

[139] Dugo G, Mondello L, Costa R, Albergamo A, Gomes T. Potential use of proteomics in shellfish aquaculture: From assessment of environmental toxicity to evaluation of seafood quality and safety. Current Organic Chemistry. 2017;**21**:402-425. DOI: 10.2174/13852728 20666161102121232

[140] Paul D, Kumar A, Gajbhiye A, Santra MK, Srikanth R. Mass spectrometry-based proteomics in molecular diagnostics: Discovery of cancer biomarkers using tissue culture. BioMed Research International. 2013;**2013**:783131. DOI: 10.1155/2013/783131

[141] Xu H, Jing L, Li G-S. Proteomic analysis of osteoblasts exposed to fluoride in vitro. Biological Trace Element Research. 2008;**123**:91-97. DOI: 10.1007/s12011-007-8086-3

[142] Kobayashi CAN, Leite A de L, da Silva TL, dos Santos LD, Nogueira FCS, Santos KS, et al. Proteomic analysis of urine in rats chronically exposed to fluoride. Journal of Biochemical and Molecular Toxicology. 2011;**25**:8-14. DOI: 10.1002/jbt.20353

[143] Carvalho JG, Leite A de L, Peres-Buzalaf C, Salvato F, Labate CA, Everett ET, et al. Renal proteome in mice with different susceptibilities to fluorosis.

PLoS One. 2013;**8**:e53261. DOI: 10.1371/journal.pone.0053261

[144] Leite AL, Lobo JGVM, da Silva Pereira HAB, Fernandes MS, Martini T, Zucki F, et al. Proteomic analysis of gastrocnemius muscle in rats with streptozotocin-induced diabetes and chronically exposed to fluoride. PloS One. 2014;**9**:e106646. DOI: 10.1371/journal.pone.0106646

[145] Kobayashi CAN, Leite AL, Silva TL, Santos LD, Nogueira FCS, Oliveira RC, et al. Proteomic analysis of kidney in rats chronically exposed to fluoride. Chemico-Biological Interactions. 2009;**180**:305-311. DOI: 10.1016/j.cbi.2009.03.009

[146] Pan Y, Lü P, Yin L, Chen K, He Y. Effect of fluoride on the proteomic profile of the hippocampus in rats. Zeitschrift fur Naturforschung. C, Journal of Bbiosciences. 2015;**70**:151-157. DOI: 10.1515/znc-2014-4158

[147] Wei Y, Zeng B, Zhang H, Chen C, Wu Y, Wang N, et al. iTRAQ-based proteomics analysis of serum proteins in Wistar rats treated with sodium fluoride: Insight into the potential mechanism and candidate biomarkers of fluorosis. International Journal of Molecular Sciences. 2016;**17**:E1644. DOI:10.3390/ij-ms17101644

[148] Sakagami H, Sugimoto M, Tanaka S, et al. Metabolomic profiling of sodium fluoride-induced cytotoxicity in an oral squamous cell carcinoma cell line. Metabolomics. 2014;**10**:270-279. DOI: https://doi.org/10.1007/s11306-013-0576-z

[149] Ministry of Health & Family Welfare Government of India. National Programme for Prevention and Control of Fluorosis (NPPCF) [Internet]. 2014. Available from: http://cghealth.nic.in/ehealth/2017/Instructions/NPPCFnewguidelinebyGOI.pdf

[150] Darchen A, Sivasankar V, Mamba BB, Narayanasamy R. Treatment of fluorosis disease and prevention of negative effects of fluoride ingestion. In: Sivasankar V, editor. Surface Modified Carbons as Scavengers for Fluoride from Water. Cham: Springer; 2016. pp. 197-210. DOI: 10.1007/978-3-319-40686-2_10

[151] Sherwood IA. Fluorosis varied treatment options. Journal of Conservative Dentistry. 2010;**13**(1):47. DOI: 10.4103/0972-0707.62631

[152] Abanto AJ, Rezende KM, Marocho SM, Cellerti P, Ciamponi AL. Dental fluorosis: Exposure, prevention and management. Medicina Oral, Patología Oral y Cirugía Bucal. 2009;**14**(2):E103-E107

[153] Levy SM. Review of fluoride exposures and ingestion. Community Dentistry and Oral Epidemiology. 1994;**22**(3):173-180. DOI: 10.1111/j.1600-0528.1994.tb01836.x

[154] Ministry of Health of the People's Republic of China. Standard of Population Total Fluoride Intake (WS/T 87-2016) [Internet]. 2016. Available from: http://www.nhfpc.gov.cn/fzs/s7852d/201606/94e135cb57b1495bb1a16f631d2265ae.shtml

Section 2

Pollution Control Approaches

Chapter 4

Polymer Hydrogels for Wastewater Treatment

Luqman Ali Shah and Sher Ali Khan

Abstract

The pollution of water resources turns into a worldwide problem because of the indiscriminate disposal of pollutants both organic and inorganic in nature. It stays hard to manner or control the purification of wastewater before it flows to water reservoirs. The growing interest in the improvement and application of novel hydrogels in wastewater remedy is due to its particular chemical characteristics along with hydrophilicity, sensitivity, and functionality. Hydrogels exhibit superior overall performance inside the adsorptive removal of a wide variety of aqueous pollutants along with heavy metals, nutrients, and toxic dyes. In this chapter, we are focusing on the behavior and importance of the hydrogels used so far for the removal of both organic and inorganic pollutants from wastewater. With this contribution, we will be able to elaborate the answer for why these hydrogels are superior than other materials used for the same purpose. More attention is given to the removal of heavy metal ions from wastewater using different hydrogel systems.

Keywords: hydrogels, hydrophilicity, sensitivity, heavy metal ions, water treatment

1. Introduction

In recent times, the fast growth of industries has caused critical troubles within the natural environment. The effluents of many industries that include paint industries, metal plating, food industries, pharmaceutical industries, and battery production, which comprise heavy metallic ions, dyes, and organic materials, are discharged without delay into water bodies and cause water pollution. These pollutants above the permissible limit cause serious effects on human beings and other terrestrial and aquatic animals. These substances penetrate and accumulate inside the bodies through food chains [1]. For the remediation and purification of waste-contaminated water, a number of different strategies were used, which include chemical precipitation [2], ion exchange [3], biological methods [4], membrane separation [5], reverse osmosis [6], coagulation and flocculation [7], catalysis [8–11], photodegradation [12], and adsorption [13, 14]. Among these strategies, adsorption is considered a cheap, quick, and environmental friendly process for wastewater treatment.

Generally, the adsorption process is broadly categorized in chemisorption and physisorption. Chemisorption also called chemical adsorption involves the formation of a chemical bond between adsorbate and adsorbent and therefore behaves as an irreversible system. Physisorption or physical adsorption takes place through physical interaction like hydrogen bonding, van der Waals, and hydrophobic interactions between adsorbate and adsorbent and acts in a reversible manner. Physical conditions such as pH, ionic energy, adsorbate and adsorbent dosage, contact time, and temperature are the most vital factors that affect the adsorption capability of hydrogels. The optimization of these factors is very crucial and should be considered first to layout the adsorption process at large scale [15, 16].

Hydrogels are three-dimensional, cross-linked, and flexible polymer networks having hydrophilic groups like hydroxyl, carboxyl, and amide [17] that swell in water and biological fluids. This swelling property is reversible and depends strongly upon environmental conditions, and therefore, these materials are also called intelligent or smart materials. Polymer hydrogels can be synthesized by different methods including freeze-drying [18], pyrogenation [19], microemulsion formation [20], and segment separation [21], having different geometries in the form of bead [22], film [23, 24], ring [25], and hole fiber depending on its application. Hydrogels respond to environmental stimuli, which include light [26], temperature [9, 11, 27], electric powered field [28], magnetic field [29], pH [30], ionic strength [31], chemical species [32], etc.

A number of efforts had been made within the field of hydrogels for wastewater treatment. In most studies, the researchers focus on the removal capability of hydrogels toward organic toxic dyes and inorganic toxic heavy metal ions, and nowadays, special interest is given to emergent pollutants. The distinguished emergent pollutants are pharmaceuticals, drugs, insecticides, pesticides, and other toxic chemical substances [33]. These pollutants even at very low concentrations in wastewater are highly dangerous to human bodies and aquatic animals [34].

2. Graphene oxide (GO)-based hydrogels for wastewater treatment

Polymer hydrogels are used for purification purpose, but due to weak mechanical strength, the use of these materials is restricted to some specific conditions. Therefore, to increase the application window for hydrogels, GO or some other inorganic species were introduced to fabricate composite hydrogels having enhanced mechanical strength. Besides the mechanical strength, the sheets of GO show excellent adsorption capacity for the elimination of toxic organic dyes from an aqueous environment. Currently, Guo et al. [35] carried out the facile synthesis of GO/polyethylenimine (PEI) by incorporation of GO in polyethylenimine (PEI) network producing a green adsorbent GO/PEI having enhanced removal capacity toward organic dyes. The hydrogen bonding and electrostatic interactions among amine groups of PEI and GO sheets accomplished the GO/PEI hydrogels. The removal performance was studied for both methylene blue (MB) and rhodamine B (RhB). The as-prepared hydrogels show complete removal of these dyes within 4 h followed by the pseudo-second-order kinetics. The superior dye removal ability of hydrogels is strongly attributed to the GO sheets, while the PEI is responsible for the facilitation of the gelation process of GO sheets. The beauty of these hydrogels is that it can be recovered and reused again without any trouble from an aqueous environment, suggesting the potential importance of these materials for wastewater treatment.

Figure 1 shows the various steps occurring in the formulation of GO/PEI hydrogels. The GO sheets having large hydrophilic functional groups, e.g., carboxyl, hydroxyl, and epoxides (**Figure 1A**), on the surface can generate hydrogen bonds with amine groups of PEI under appropriate conditions. Consequently, PEI (**Figure 1B**) facilitates the gelation of GO sheets in an aqueous solution and also reveals the correct adsorption and adhesion properties. It was found that the dye adsorption capacity of the GO/PEI hydrogels increased with the amount of PEI in polymer network. It is due to the electrostatic attractions among the amine

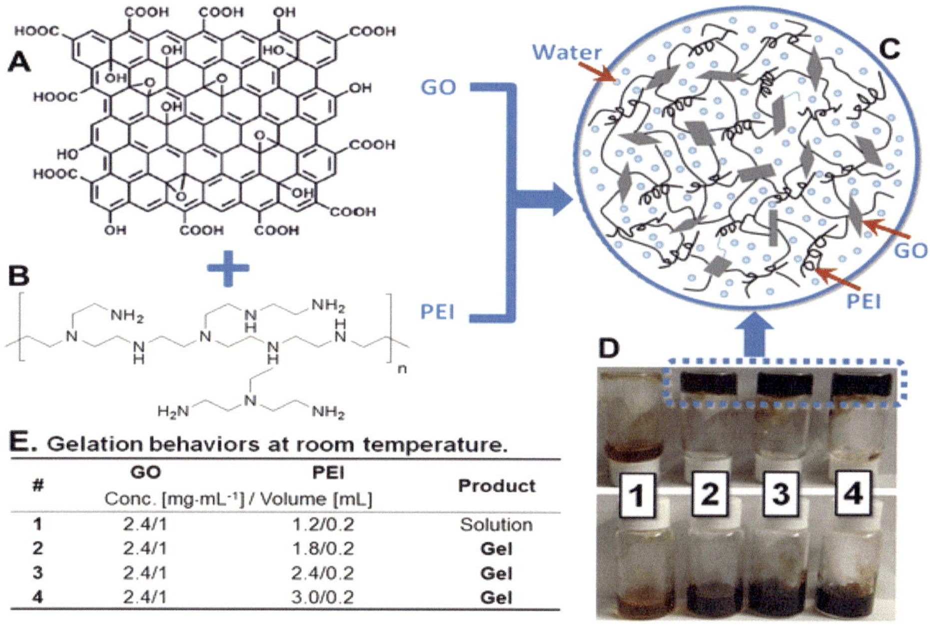

#	GO	PEI	Product
	Conc. [mg·mL⁻¹] / Volume [mL]		
1	2.4/1	1.2/0.2	Solution
2	2.4/1	1.8/0.2	Gel
3	2.4/1	2.4/0.2	Gel
4	2.4/1	3.0/0.2	Gel

Figure 1.
Schematic representation of the formulation of GO/PEI hydrogel. (A) GO, (B) PEI, (C) GO/PEI hydrogels, and (D) and (E) are the gelation pictures [35].

functionalities in polymer network and the dye molecules. From these results, we can conclude that the adsorption capability of the GO/PEI hydrogels is largely attributed to PEI, and GO increases the mechanical strength. Therefore, in maximum composite substances, PEI is extensively applied as a robust chelating agent and organic intermediate. Furthermore, the GO/PEI composite hydrogel showed very stable self-assembly behaviors by using hydrogen bonding and electronic interactions, which confirmed an extraordinary possibility to launch PEI to motive secondary waste as dye adsorbents for wastewater treatment.

3. Jute/polyacrylic acid hydrogel systems for wastewater treatment

In real life the materials having high adsorption capacity, rapid removal kinetics, reusability, and cost-effective are preferred to utilize in wastewater treatment. To achieve these properties, a porous Jute/Polyacrylic acid (Jute/PAA) hydrogel was prepared. The high permeability and 80 wt% water in polymer network of Jute/PAA hydrogel made the inner sites fully available for the adsorption of metal ions. The Jute/PAA gel adsorbs heavy metal ions particularly Cd^{2+} and Pb^{2+} from wastewater with very high adsorption capacities of 401.7 and 542.9 mg/g for Cd^{2+} and Pb^{2+}, respectively. Furthermore, the adsorption equilibrium was reached within 10 min for 40 mg/L of initial ion concentration using 1 g/L of hydrogel. Meanwhile, the elimination efficiencies reached 81% for Pb^{2+} and 79.3% for Cd^{2+}. The materials were checked for other metal ions such as Cu, Zn, Mn, Cr, and Fe in melting wastewater under the same environmental conditions using different amount of hydrogel, and the results are tabulated in **Table 1**. The concentrations of Pb, Cd, and Cr reduced beneath 0.001 mg/L with the use of 4 g/L adsorbent. In the fixed-bed column experiments, the treatment quantity of melting wastewater reached 2900 BV (32.8 L) only generating 50 BV (565 ml) eluent. This study strongly helps in the development of a realistic adsorption system based on hydrogel adsorbents

Sorbent dosage (g/L)	Metal ion concentration (mg/L)						
	Pb	Cd	Cu	Zn	Mn	Cr	Fe
0	3.825	6.075	9.325	188.6	17.05	0.25	9.75
1	0.725	1.255	1.550	132.5	13.25	<0.001	<0.01
2	<0.001	0.225	0.350	94.65	11.15	<0.001	<0.01
3	<0.001	0.075	0.085	40.55	6.55	<0.001	<0.01
4	<0.001	<0.001	0.010	17.95	3.05	<0.001	<0.01

Table 1.
Metal ion concentrations of industrial effluent by different adsorbent dose [37].

for the wastewater treatment. Therefore, the removal performance of hydrogels toward heavy metal ions was investigated in real industrial water collected from smelting plant.

The preferential removal to low degree of Fe and Cr is because of the excessive average valence electron power and the configuration of $3d^6 4s^2$ subshell, which provides empty orbital and strongly coordination ability [36]. When increasing the adsorbent dosage to 2 g/L, Pb was preferentially removed in divalent metallic ions with the residual attention below 0.001 mg/L, probably because of the better electronegativity of Pb. While in addition to increasing the Jute/PAA gel dosage to 4 g/L, the Cd ions could be completely adsorbed, and the removal efficiencies of Cu, Zn, and Mn ions attain up to 99.8, 90.5, and 61.6%, respectively. From the obtained results, it is confirmed that the Jute/PAA hydrogel has a strong capability in the removal of heavy metal ions from commercial effluents.

Table 1 shows the adsorption statistics of heavy metal ions through different amount of Jute/PAA hydrogel dosage after remedy for 2 h.

4. Carboxymethyl cellulose/2-acrylamido-2-methyl propane sulfonic acid hydrogels

A series of functional copolymer hydrogels composed of carboxymethyl cellulose (CMC) and 2-acrylamido-2-methyl propane sulfonic acid (AMPS) have been synthesized using γ-radiations and prompted copolymerization and cross-linking, and their swelling ability was investigated to optimum conditions. The capacity of these hydrogels was tested in the recovery of toxic heavy metal ions, i.e., Mn^{+2}, Co^{+2}, Cu^{+2}, and Fe^{+3}, from their aqueous solutions. The hydrogels showed a pronounced effect on the removal of metal ions. The pronounced removal ability is due to the existence of AMPS in the inner composition of hydrogels, which has a strong chelating potential and forms a stable interaction with metal ions. Therefore, by increasing the AMPS concentration in polymer chains, the chelating potential increases, and the hydrogels will show enhanced removal performance. The prepared hydrogels of CMC/AMPS were stable and utilized in a multiple cycles with no reduction compared to their initial performance.

The adsorption process will be more active and favorable if the interaction of metal ions with the adsorbent is strong. Therefore, the effect of contact time on the adsorption ability of the CMC/AMPS copolymer hydrogel toward metal ions, i.e., Co, Mn, Cu, and Fe, changed. Initially the adsorbed amount of metal ions was efficient and then reduced. Few researchers studied and found that the decrease of chelating ability resulted in polymer chain shrinkage that takes area due to adsorption happening at the hydrogel network, due to which the diffusion of cations

Metal	K (×10^{-3} s^{-1})	(C_e/C_0)	R^2	K_d (s^{-1})	K_s (s^{-1})
Fe	5.24	0.842	0.994	1.176	0.013
Cu	6.14	0.884	0.986	1.064	0.013
Mn	4.94	0.847	0.994	0.400	0.008
Co	4.99	0.910	0.989	0.719	0.011

Table 2.
Adsorption rate constant of different metals on CAP6 hydrogel [38].

become difficult inside the bulk of the hydrogel. The fast adsorption process in the initial stage occurs on the surface and after the adsorption takes place inside the hydrogel network and slows down due to the penetration of metal ions through the pores in the hydrogel matrix. The sorption process depends on the intraparticle diffusion, chelation, and ion interactions. **Table 2** shows the adsorption rate constant obtained by removing different metal ions from wastewater.

In other studies, the environment friendly carboxymethyl cellulose (CMC) hydrogel beads were successfully prepared using epichlorohydrin (ECH) as a cross-linking agent through ether linkage formed between ECH and CMC in the suspension of fluid wax. The characteristic bands in FTIR spectra confirmed the ether linkage. The prepared hydrogel beads were 4 mm in diameter with fully transparent and apparently spherical geometry. It was further confirmed by X-ray diffraction (XRD) patterns that the adsorption of metal ion onto the oxygen atom of carboxyl group changed the crystallinity of hydrogels. The adsorption capacity depends on the initial concentrations of metal ions and the pH value of metal ion solution and was found increased with increase in pH and initial concentration of metal ion solution. After the application of Freundlich and Langmuir isotherm models on the data obtained from the batch adsorption experiments, it was found that the sorption mechanism of the hydrogel beads for metal ions follows the Langmuir model. The maximum adsorption values of hydrogel beads for metal ions is 6.49, 4.06, and 5.15 mmol/g for Cu^{2+}, Ni^{2+}, and Pb^{2+}, respectively.

Superabsorbent hydrogel beads based on CMC were prepared by suspension cross-linking method and were characterized by FTIR, XRD, and SEM. The crystallinity of these hydrogels was less than the pure CMC and was confirmed by XRD analysis. It was assumed that the adsorption of metal ions on hydrogel beads formed coordination bonds with the oxygen atoms in the carboxyl groups of hydrogel beads and showed good adsorption ability for heavy metal ions. The maximum amount of adsorbed metal ions from the data of Langmuir model is 6.49, 4.06, and 5.15 mmol/g for Cu^{2+}, Ni^{2+}, and Pb^{2+}, respectively, at pH 7. The study of adsorption indicates that the hydrogel beads have a potential application and can be applied on large scale for wastewater treatment [39].

5. Hydrogels based on natural polysaccharides for wastewater treatment

Nowadays hydrogels based on bio-originated polymers like chitosan, maltodextrin, and gum arabic with and without magnetite nanoparticles had been employed as adsorbents for entrapping of heavy metal ions from aqueous solutions. The adsorption and removal of heavy metal ions by hydrogels are due to the diffusion of water molecules inside the hydrogel network and confirmed by adsorption kinetics using the Fickian equation. The shape of macromolecules rests and can be laid low with the presence of magnetite nanoparticles, and this effect is associated with the reticulation factors inside the hydrogel network. The CS-, M-malto-, and

M-GA-based hydrogels were applied for the removal of Cd^{2+} ions from aqueous solutions under the controlled conditions of pH 4.5–5.5, initial concentration of 20 mg L^{-1}, and dried hydrogel mass of 100 mg [40]. By applying three adsorption isotherms, i.e., Langmuir, Freundlich, and Redlich-Peterson, it was found that a change in physiochemical phenomena related to Cd^{2+} adsorption occurred and the data fitted to the Langmuir or Redlich-Peterson models more than the Freundlich model. The beauty and advantage of these materials are recovery by a simple magnetic field compared to other hydrogels that require mostly ultracentrifugation or using solvents like HCl, HNO_3, etc.

The removal of heavy metals from water and industrial effluents has been the goal of a large number of studies. Paulino et al. concluded from their results that hydrogels based on polysaccharides such as CS, M-malto, and M-GA are important absorbers for the treatment of wastewater and removal of heavy metal ions from industrial effluents. It was also elaborated that the diffusion of Cd^{2+} through polymer hydrogel network changed when magnetic nanoparticles were introduced into polymer network. Based on Fickian parameters, the hydrogels have diffusion properties with a tendency toward macromolecular relaxation, which is very important for adsorption studies of both organic and inorganic pollutants. Different parameters such as contact time, pH, initial hydrogel dosage, and initial concentration of the Cd^{2+} solution were studied to examine the potential application of hydrogels with and without magnetic properties for the removal of Cd^{2+} from water and effluents. The results confirmed that hydrogels without magnetic nanoparticles based only on CS, M-malto, and M-GA can be applied more efficiently in wastewater treatment for the removal of Cd^{2+} compared to hydrogels with magnetic particles. However, the regeneration of magnetic hydrogels can be done more easily with the application of magnetic field which is environmental friendly and green approach [40].

6. Treatment of polluted water resources using reactive polymeric hydrogel

An experimental work becomes performed to study the overall performance of the prepared polyvinyl pyrrolidone/acrylic acid (PVP/AAc) copolymer hydrogel to chelate heavy metals from the bulk material [41]. The results clearly indicate that PVP/AAc copolymer hydrogel has excessive binding capacities and proper adsorption kinetics for the metal ions. The sorption of these metal ions follows the Langmuir adsorption isotherm. The feasibility for the uses of PVP/AAc hydrogel for the treatment of polluted samples, accrued from distinct water assets in Helwan location (Egypt), was investigated. The results showed that by using these hydrogels, we can obtain pure usable water from wastewater.

In recent years, there has been considerable interest in the chelation of metal ions by insoluble cross-linked polymeric substrate. Such substrates have advantages over soluble materials of easy separation from the reaction medium, leading to operational flexibility of their facial regenerability and of higher stability. The removal performance of hydrogels can be disturbed in the presence of other metal salts like NaCl, $MgCl_2$, $CaCl_2$, etc. in polluted water and affect the chelating ability of the species. This effect was studied by Shawky et al. in the adsorption of Fe ions by PVP/AAc hydrogel in the presence of different metal salts, and the effect is shown in **Figure 2**. No effect on the adsorption of Fe ions was observed in case of NaCl even at high concentration of NaCl. Furthermore, Fe adsorption is not suffering by low concentrations of $CaCl_2$ and $MgCl_2$; however, at higher concentrations, the adsorption decreases, and this could be attributed to the affinity of the reactive polymers toward alkaline earth metals as compared to transition metal ions.

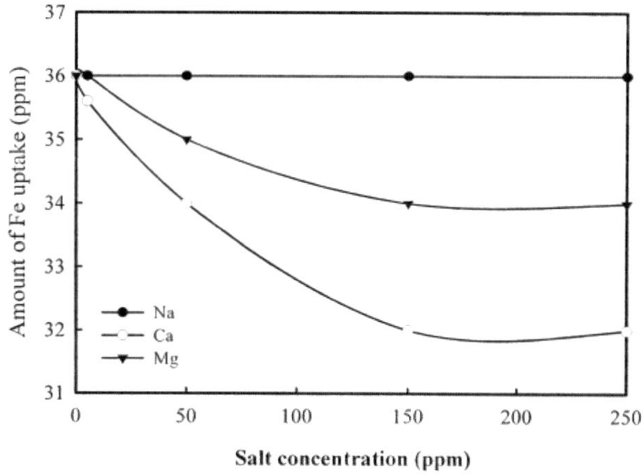

Figure 2.
Effect of different concentrations of dissolved salts on Fe uptake [41].

The results obtained clearly demonstrate and confirmed the applicability of PVP/AAc hydrogel for wastewater treatment. Water resources in the Helwan area showed that trace metal contents are very high when the analysis of nine water samples was carried out. The hydrogel treatment resulted in a satisfactory removal of polluted heavy metals especially iron, manganese, and aluminum [41].

7. Conclusion

The smart polymer hydrogels prepared from both synthetic and natural polymers can be used successfully with full confidence for wastewater treatment. However, the properties of these hydrogels will be kept according to the required environmental conditions by changing the composition of polymer networks. The performance of few polymer hydrogels was explained and executed in this chapter, which clearly indicates that due to smart behavior, easy synthesis, recycling, low cost, environment friendly, biocompatibility, etc. make these hydrogels as efficient candidate compared to other materials for wastewater treatment. By reading this chapter, the researchers could find new approaches which help them in designing new hydrogel systems for different applications.

Acknowledgements

The authors are very grateful to acknowledge the Higher Education Commission of Pakistan for financial support under Grant No: 7309/KPK/NRPU/R and D/HEC/2017.

Author details

Luqman Ali Shah* and Sher Ali Khan
Polymer Laboratory, National Centre of Excellence in Physical Chemistry, University of Peshawar, Pakistan

*Address all correspondence to: luqman_alisha@yahoo.com; luqman_alisha@uop.edu.pk

References

[1] Fu F, Wang Q. Removal of heavy metal ions from wastewaters: A review. Journal of Environmental Management. 2011;**92**(3):407-418

[2] Charerntanyarak L. Heavy metals removal by chemical coagulation and precipitation. Water Science and Technology. 1999;**39**(10-11):135-138

[3] Wang M et al. Hybrid algal photosynthesis and ion exchange (HAPIX) process for high ammonium strength wastewater treatment. Water Research. 2018;**142**:65-74

[4] Vashi H, Iorhemen OT, Tay JH. Aerobic granulation: A recent development on the biological treatment of pulp and paper wastewater. Environmental Technology and Innovation. 2018;**9**:265-274

[5] Cui X et al. Dark matter results from 54-ton-day exposure of PandaX-II experiment. Physical Review Letters. 2017;**119**(18):181302

[6] Volpin F et al. Simultaneous phosphorous and nitrogen recovery from source-separated urine: A novel application for fertiliser drawn forward osmosis. Chemosphere. 2018;**203**:482-489

[7] Choumane FZ et al. Valorisation of a bioflocculant and hydroxyapatites as coagulation-flocculation adjuvants in wastewater treatment of the steppe in the wilaya of Saida (Algeria). Ecological Engineering. 2017;**107**:152-159

[8] Shah LA et al. Synthesis of sensitive hybrid polymer microgels for catalytic reduction of organic pollutants. Journal of Environmental Chemical Engineering. 2016;**4**(3):3492-3497

[9] Shah LA et al. Thermal and pH dual responsive copolymer and silver nanoparticle composite for catalytic application. Chinese Journal of Chemistry. 2015;**33**(4):467-472

[10] Shah LA et al. Silver nanoparticles fabricated hybrid microgels for optical and catalytic study. Journal of the Chemical Society of Pakistan. 2016;**38**(5):850-858

[11] Shah LA et al. Ag-loaded thermo-sensitive composite microgels for enhanced catalytic reduction of methylene blue. Nanotechnology for Environmental Engineering. 2017;**2**(1):14

[12] Shah LA et al. TiO_2 nanotubes doped poly (vinylidene fluoride) polymer membranes (PVDF/TNT) for efficient photocatalytic degradation of brilliant green dye. Journal of Environmental Chemical Engineering. 2019;7:103291

[13] Javed R et al. Uptake of heavy metal ions from aqueous media by hydrogels and their conversion to nanoparticles for generation of a catalyst system: Two-fold application study. RSC Advances. 2018;**8**(27):14787-14797

[14] Li J et al. Metal–organic framework-based materials: Superior adsorbents for the capture of toxic and radioactive metal ions. Chemical Society Reviews. 2018;**47**(7):2322-2356

[15] Rehman TU et al. Zwitterionic superabsorbent polymer hydrogels for efficient and selective removal of organic dyes. RSC Advances. 2019;**9**(32):18565-18577

[16] Shah LA. Developing Ag-tercopolymer microgels for the catalytic reduction of p-nitrophenol and EosinY throughout the entire pH range. Journal of Molecular Liquids. 2019;**288**:111045

[17] Cruz H et al. Rapid removal of ammonium from domestic wastewater

using polymer hydrogels. Scientific Reports. 2018;**8**(1):2912

[18] Butylina S, Geng S, Oksman K. Properties of as-prepared and freeze-dried hydrogels made from poly (vinyl alcohol) and cellulose nanocrystals using freeze-thaw technique. European Polymer Journal. 2016;**81**:386-396

[19] Badiger MV, McNeill ME, Graham NB. Porogens in the preparation of microporous hydrogels based on poly (ethylene oxides). Biomaterials. 1993;**14**(14):1059-1063

[20] Ghayempour S, Montazer M. A modified microemulsion method for fabrication of hydrogel Tragacanth nanofibers. International Journal of Biological Macromolecules. 2018;**115**:317-323

[21] Omidian H, Rocca JG, Park K. Advances in superporous hydrogels. Journal of Controlled Release. 2005;**102**(1):3-12

[22] Farhoudian S, Yadollahi M, Namazi H. Facile synthesis of antibacterial chitosan/CuO bio-nanocomposite hydrogel beads. International Journal of Biological Macromolecules. 2016;**82**:837-843

[23] Zhang W et al. A family of metal-organic frameworks exhibiting size-selective catalysis with encapsulated noble-metal nanoparticles. Advanced Materials. 2014;**26**(24):4056-4060

[24] Zhang M, Zhang Y, Helleur R. Selective adsorption of Ag+ by ion-imprinted O-carboxymethyl chitosan beads grafted with thiourea–glutaraldehyde. Chemical Engineering Journal. 2015;**264**:56-65

[25] Deng Y et al. Insight into highly efficient simultaneous photocatalytic removal of Cr (VI) and 2, 4-diclorophenol under visible light irradiation by phosphorus doped porous ultrathin g-C3N4 nanosheets from aqueous media: Performance and reaction mechanism. Applied Catalysis B: Environmental. 2017;**203**:343-354

[26] Tomatsu I, Peng K, Kros A. Photoresponsive hydrogels for biomedical applications. Advanced Drug Delivery Reviews. 2011;**63**(14-15):1257-1266

[27] Shah LA, Sayed M, Siddiq M. Fabrication of Ag and Au nanoparticles in cross-linked polymer microgels for their comparative catalytic study. Materials Science-Poland. 2017;**35**(3):651-659

[28] Kim SJ et al. Behavior in electric fields of smart hydrogels with potential application as bio-inspired actuators. Smart Materials and Structures. 2005;**14**(4):511

[29] Liu TY et al. Magnetic-sensitive behavior of intelligent ferrogels for controlled release of drug. Langmuir. 2006;**22**(14):5974-5978

[30] Rogina A et al. Cellular hydrogels based on pH-responsive chitosan-hydroxyapatite system. Carbohydrate Polymers. 2017;**166**:173-182

[31] Zhang R et al. A novel pH-and ionic-strength-sensitive carboxy methyl dextran hydrogel. Biomaterials. 2005;**26**(22):4677-4683

[32] Yoshida R, Okano T. Stimuli-responsive hydrogels and their application to functional materials. In: Biomedical Applications of Hydrogels Handbook. New York, Dordrecht, Heidelberg, London: Springer; 2010. pp. 19-43

[33] Geissen V et al. Emerging pollutants in the environment: A challenge for water resource management. International Soil and Water Conservation Research. 2015;**3**(1):57-65

[34] Pakdel PM, Peighambardoust SJ. Review on recent progress in chitosan-based hydrogels for wastewater treatment application. Carbohydrate Polymers. 2018;**201**:264-279

[35] Guo H et al. Preparation of graphene oxide-based hydrogels as efficient dye adsorbents for wastewater treatment. Nanoscale Research Letters. 2015;**10**(1):272

[36] Berlia R, Kumar MP, Srivastava C. Electrochemical behavior of Sn–graphene composite coating. RSC Advances. 2015;**5**(87):71413-71418

[37] Zhou G et al. Efficient heavy metal removal from industrial melting effluent using fixed-bed process based on porous hydrogel adsorbents. Water Research. 2018;**131**:246-254

[38] Ahmed S et al. Advances in heterogeneous photocatalytic degradation of phenols and dyes in wastewater: A review. Water, Air, and Soil Pollution. 2011;**215**(1-4):3-29

[39] Yang S et al. Hydrogel beads based on carboxymethyl cellulose for removal heavy metal ions. Journal of Applied Polymer Science. 2011;**119**(2):1204-1210

[40] Paulino AT et al. Efficiency of hydrogels based on natural polysaccharides in the removal of Cd^{2+} ions from aqueous solutions. Chemical Engineering Journal. 2011;**168**(1):68-76

[41] Shawky H et al. Treatment of polluted water resources using reactive polymeric hydrogel. Journal of Applied Polymer Science. 2006;**100**(5):3966-3973

Chapter 5

Biological Remediation of Phenoxy Herbicide-Contaminated Environments

Magdalena Urbaniak and Elżbieta Mierzejewska

Abstract

Phenoxy herbicides such as 2,4-dichlorophenoxyacetic acid (2,4-D) and 2-methyl-4-chlorophenoxyacetic acid (MCPA) are widely used in agriculture to control broadleaf weeds. Although their application has helped to increase the yield and value of crops, they are also recognized as a source of emerging environmental contamination. Their extensive use may promote contamination of soil, surface, and groundwater and lead to increased inhibition of plant development and soil toxicity. Hence, there is an urgent need to identify nature-based methods based on appropriate biological remediation techniques, such as bio-, phyto-, and rhizoremediation, that enable the effective elimination of phenoxy herbicides from the environment. Bioremediation typically harnesses microorganisms and their ability to utilize recalcitrant contaminants in complete degradation processes, while phytoremediation is a cost-effective, environmentally friendly strategy that uses plants to transform or mineralize xenobiotics to less or nontoxic compounds. Rhizoremediation (microbe-assisted phytoremediation), in turn, is based on the interactions between plant roots, root exudates enriched in plant secondary metabolites, soil, and microorganisms. Based on the above, this chapter presents current knowledge on the properties of phenoxy herbicides, as well as the concentrations detected in the environment, their toxicity, and the biological remediation techniques used for safe removal of the compounds of interest from the environment.

Keywords: 2,4-D, MCPA, bioremediation, phytoremediation, rhizoremediation, toxicity, degradative genes

1. Phenoxy herbicides: general information

2,4-Dichlorophenoxyacetic acid (2,4-D) and 2-methyl-4-chlorophenoxyacetic acid (MCPA) are the most commonly used phenoxy acid herbicides in agriculture, and 2,4-D is now the fifth most extensively used active ingredient (a.i.) in the US agricultural and home/garden market sector [1]. In addition, in 2016, 6.5 mln kg of herbicides based on phenoxy-phytohormones (2,4-D and MCPA) were sold in in the EU, including ~2 mln kg sold in Poland [2].

Phenoxy herbicides are typically used to protect wheat, one of the most extensively cultivated crops, because they selectively control the growth of dicotyledonous weeds [3]. They are applied as post emergence agents and taken up by broad-leaved plants. 2,4-D has also been extensively used as an anti-stalling agent for the postharvest fresh fruit industry [4].

These herbicides are based on ring-like structures and have at least one chlorine atom attached to the ring at different positions [5]. Their action is similar to that of phytohormones (auxins) insofar that they can redirect the regulation of plant growth/physiological processes, resulting in nutrition deficiency and subsequent plant death [6].

They are typically released to the environment in the form of commercial products containing phenoxy acids salts or esters; however, they immediately hydrolyze to their corresponding anionic or neutral form [7]. The dosage of phenoxy herbicides lies in the range of 0.8–1.8 kg of a.i. per ha. Their transport through the environment is governed by soil and climate factors (e.g., distribution of soil particles, soil permeability, soil depth, soil pH, soil organic matter content, land slope) [8], and their retention and translocation in the soil profile also depend on their chemical and physical properties, which are described by several parameters (**Table 1**), particularly pK_a (acid dissociation constant), logP (octanol-water partition coefficient), and K_{oc} (organic carbon distribution coefficient). The degree of adsorption and desorption depends on time and the physicochemical properties of soil; however, 2,4-D and MCPA are rather poorly adsorbed on the soil particles in comparison to their derivatives, which have different sorption characteristics [7].

Although phenoxy herbicides are described as nonpersistent and weakly adsorbed ($K_{oc} < 50$) in soil, they can be transported with runoff and in soil profile and reach terrestrial and water ecosystems (surface and groundwater). **Figure 1** summarizes the transport and transfer processes of phenoxy herbicides in the environment. After they are applied to land, they are spread through several processes, including sorption/desorption, leaching, runoff, and plant uptake [8]. Phenoxy herbicide molecules are negatively charged and are therefore highly mobile at neutral pH. In groundwater, they are nonvolatile and persistent to hydrolysis, but they can be degraded biologically under both aerobic and anaerobic conditions. These herbicides demonstrate significantly greater persistence in temperate climates characterized by low winter temperatures and, in many regions, by depleted soil organic carbon content and acidic pH [7].

Properties	2,4 D	MCPA
Chemical structure	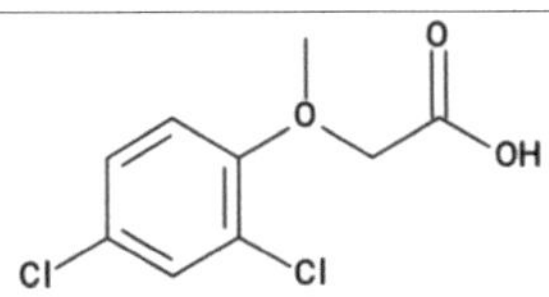	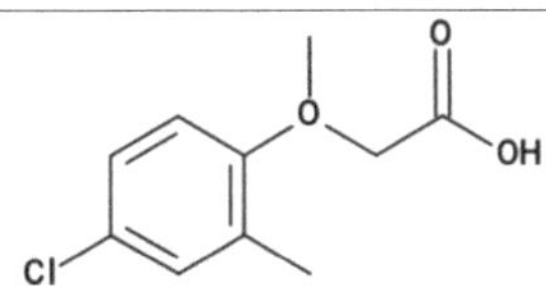
IUPAC name	(2,4-Dichlorophenoxy)acetic acid	(4-Chloro-2-methylphenoxy)acetic acid
Molar mass (g/mol)	220.04	200.62
Solubility in H_2O (mg/L) [7]	450	720
pK_a [7]	2.73–2.87	3.73
K_{oc} [7]	20.56 mL/g	25–157 mL/g
logP [7]	2.50/2.58 ± 0.36	2.41/2.49 ± 0.27
t1/2 in water [9, 10]	1–14 days	15 days
t1/2 in soil [9, 10]	2–4 weeks	3–4 weeks

Table 1.
Physical and chemical properties of 2,4-D and MCPA.

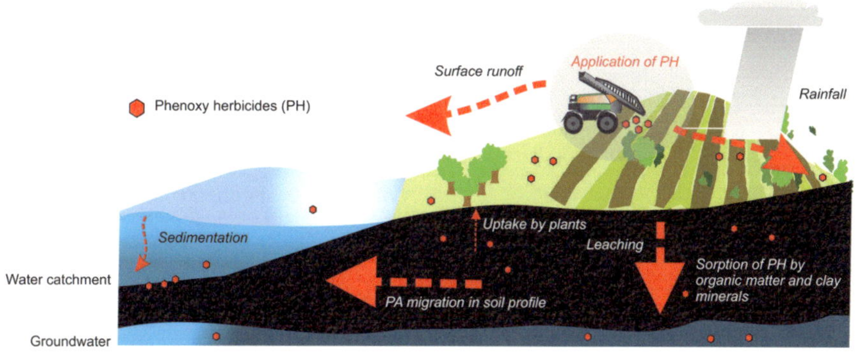

Figure 1.
Transport and transfer processes of phenoxy herbicides in the environment.

2. Phenoxy herbicides: potential contaminants of soil and water environments

Extensive use of phenoxy herbicides can threaten surface and groundwater ecosystems by promoting the contamination of soil matrices. The International Agency for Research on Cancer classifies phenoxy acids as "possibly carcinogenic to humans." Gupta et al. [11] report that 0.5 kg/ha is the optimal concentration of 2,4-D which avoids contamination of environmental matrices, with the effect of higher concentrations of 2,4-D on the environment being dependent on irrigation treatment. Hence, little is known of the distribution of phenoxy herbicides in the environment. Data from several sources have identified increased levels of 2,4-D and MCPA in the soil, ground-, surface, and drinking water (**Table 2**); for example, Ignatowicz and Struk-Sokołowska [12] note that the concentration of phenoxy herbicides in the Narew River (Poland) fluctuated seasonally from 0 to even 150 μg/L. The concentration of MCPA in the Parramatta River (Sydney Estuary, Australia) was 0.061 μg/L; however, its presence in river water was caused by increased runoff of storm water [13]. MCPA concentration has been found to be as high as 42.40 μg/L in the Rhone River (France) [14] and to be as little as 0.58 μg/L in Brejo of Cagarrão Stream (Portugal) [15]. The 2,4-D concentration has been found to vary from 1.678 μg/L in the water of McGregor Creek (Canada) [16] to 329.42 μg/L in water from a rice field [17]. By contrast, the maximum permissible concentration of pesticide residues in drinking water is 0.50 μg/L (Directive E98/83/EC). The data presented in **Table 2** and described above indicate that phenoxy herbicides should be considered as emerging contaminant especially in water resources.

Despite the diversified levels of phenoxy herbicides noted in worldwide environments (**Table 2**), it has to be underlined that these compounds can exert serious toxic effects on the sustainability of ecosystems, even at lower concentrations (e.g., 0.275 μg/L) (**Table 3**). According to recent research, predicted no effect concentration (PNEC) for aquatic organisms is 500 μg/L for 2,4-D and 0.022 μg/L for MCPA [20]; however, PNEC has not yet been determined for terrestrial organisms.

Because the mode of action of phenoxy herbicides mimics that of plant growth hormones, their application causes disturbances among a range of physiological processes [21]. 2,4-D inhibits root/hypocotyl elongation in *Sinapis arvensis* (wild mustard) and disrupts mesophyll cell structure in *Pisum sativum* (pea) [22, 23]. There is increasing concern that 2,4-D has negative influence on water ecosystems, leading to cellular deformation of green algae, such as *Ankistrodesmus falcatus* [24];

Compound	Concentration	Environmental matrices	Source
2,4-D	1.678 μg/L	Water from McGregor Creek (Canada)	[16]
2,4-D	103.99–329.42 μg/L	Water from rice field (Malaysia)	[14]
2,4-D	0.0052 mg/kg	Soil from cereal plantations (Poland)	[15]
2,4-D	0.513 μg/L	Lebo drain	[16]
MCPA, MCPP, 2,4-D	0–150 μg/L	Water from Narew River (Poland)	[12]
MCPA	0.0046 mg/kg	Soil from cereal plantations (Poland)	[15]
MCPA	0.08–42.40 μg/L	Water from Rhône River Delta (France)	[18]
MCPA	0.58 μg/L	Water from Brejo of Cagarrão Stream (Portugal)	[17]
MCPA	0.061 μg/L	Water from Parramatta River—Sydney Estuary (Australia)	[13]
MCPA	82.75–354.28 μg/L	Water from rice field (Malaysia)	[14]
MCPA	0.002–0.010 mg/kg	Soil from potato plantation (Poland)	[19]

Table 2.
Concentration of phenoxy acids observed in various environments.

Dose of phenoxy herbicide	Exposure time	Test organism	Effect on organism	Source
220.04 μg/L 2,4-D	2 days	*Sinapis arvensis* (wild mustard)	Inhibition of root and hypocotyl elongation	[22]
5.06 mg/L 2,4-D	72 hours	*Pisum sativum* (pea)	Severe disturbances in mesophyll cell structure and proliferation of vascular tissue in young leaves	[23]
10, 100, 500, 1000 μg/L MCPA	7 days	*Hydrilla verticillata* (waterthyme)	Disturbance of growth, anatomy, and physiology	[36]
IC50 1353.80 mg/L 2,4-D	96 hours	*Ankistrodesmus falcatus* (green microalgae)	External morphological alterations	[24]
IC50 71.20 mg/L 2,4-D	96 hours	*Microcystis aeruginosa* (toxigenic cyanobacteria)	Stimulation of the production of cyanotoxins	
LC50 66 mg/L 2,4-D	96 hours	*Cyprinus carpio* (common carp)	Behavioral changes	[26]
LC50 9.06 and 7.76 mg/L 2,4-D	96 and 168 hours	*Rhinella arenarum* (species of toad)	Reduced body size, delayed development, microcephaly, agenesis of gills, abnormal cellular proliferation processes	[27]
10–500 mg/L 2,4-D	1 hour	Human erythrocytes	Hemolysis	[29]
0.275, 2.75, and 27.5 μg/L 2,4-D and MCPA	30 minutes	Hepatic cells of *Metynnis roosevelti* (species of serrasalmid fish)	Damage of cellular metabolism and homeostasis; increased oxidative stress	[21]

Table 3.
The results of toxicological tests and effects of 2,4-D and MCPA on selected organisms.

malformations and behavioral changes to various fish, including *Cyprinus carpio* (common carp) and *Danio rerio* (zebrafish) [25, 26]; abnormal cellular proliferation in amphibians such as *Rhinella arenarum* (species of toad) [27]; and the development of nonviable embryos in invertebrates, such as *Biomphalaria glabrata* (species of freshwater snail) [28]. Sarikaya and Yilmaz [26] report that 2,4-D (66,000 μg/L) causes internal hemorrhage and behavioral changes in *C. carpio*.

Among animals, phenoxy herbicide application results in the inhibition of crucial enzymes in cell metabolism, including mitochondrial enzymes and those associated with DNA synthesis (**Table 3**). 2,4-D has also been found to induce erythrocyte lysis under laboratory conditions [29]. It is interesting to note that the intermediates formed during the degradation processes of 2,4-D, such as 2,4-dichlorophenol (2,4-DCP) and 3,5-dichlorocatechol (3,5-DCC), exhibit a strong ecotoxic effect on various organisms, including *N. tabacum* cells [30]; however, Taylor et al. [31] report that 2,4-DCP toxicity was found to be less phytotoxic than 2,4-D, under both in vitro and in vivo conditions, while 3,5-DCC exhibits higher toxicity than its parent compound [32].

Several studies have revealed that MCPA also can have a negative impact on the environment: MCPA application caused up to a 56% reduction in dehydrogenase, urease, and phosphatase activities and ergosterol content in soil [32]. In addition, this application leads to increased soil phytotoxicity to *Fagopyrum esculentum* var. Kora (buckwheat) and promoted stem deformation and leaf discoloration [33]. Mierzejewska et al. [34] note that a commercial product containing MCPA was highly toxic to the monocotyledon *Sorghum saccharatum* (sorghum) and dicotyledons *Lepidium sativum* (garden cress) and *Sinapis alba* (white mustard), inducing nearly 100% root growth inhibition. The authors also note that after 3 weeks of incubation at an ambient temperature, the high initial phytotoxicity was reduced to 3% for *L. sativum* and 34% for *S. alba* and that *S. saccharatum* demonstrated a 12% stimulation of root growth in comparison to uncontaminated control soil. The negative influence of MCPA on *L. sativum*, *S. alba*, and *S. saccharatum* growth was also confirmed by Urbaniak et al. [35]. Similarly to 2,4-D, MCPA causes also negative effects on freshwater organisms such as the freshwater crustaceans *Daphnia magna*, *Thamnocephalus platyurus*, and *Artemia franciscana* and alga *Selenastrum capricornutum* [36]. Both herbicides were found to induce the action of hepatic enzymes involved in detoxification and lipid peroxidation [21].

These studies emphasize the important role played by ecotoxicological approaches in evaluating the effect of chemical stressors observed in the ecosystem communities. Despite the relatively short half-life (**Table 1**) of 2,4-D and MCPA, their remnants can be transported and deposited extensively in the environment, and this can present a potential threat to the soil and water ecosystems as well as to human health. Therefore, there is a need to identify nature-based solutions such as bio-, phyto-, and rhizoremediation that can enhance the process of phenoxy herbicide elimination from the environment.

3. Phenoxy herbicides: removal using biological methods

One approach to removing phenoxy herbicides (2,4-D and MCPA) from soil is via degradation by the soil microbiota (biodegradation). This is achieved most effectively by bacteria harboring the appropriate functional genes, which are involved in the phenoxy herbicide degradation pathways (**Figure 2**). Alternatively, plants can be used to decontaminate sites, a process known as phytoremediation (**Figure 2**). Another promising approach, rhizoremediation, enhances the removal

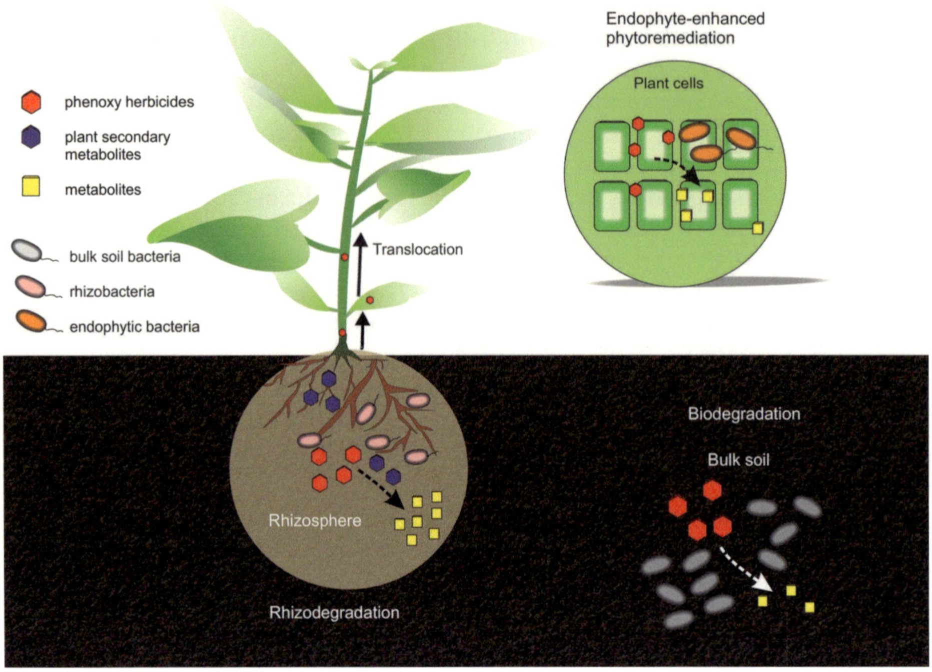

Figure 2.
The biological processes of phenoxy herbicide biodegradation mediated by soil, rhizospheric, and endophytic bacteria.

of such recalcitrant xenobiotics from the environment by exploiting the interactions between selected plants (able to grow under the presence of given xenobiotics such as phenoxy herbicides), root exudates (including plant secondary metabolites, PSMs), and microorganisms (**Figure 2**). The purpose of this section is to review the literature on established and potential biological methods of phenoxy herbicide removal from environmental matrices.

3.1 Bioremediation

Bioremediation is a method that uses microbiological processes to degrade or transform contaminants to less toxic or nontoxic forms. Biodegradation of organic contaminants occurs very slowly in bulk soil; therefore biostimulation and bioaugmentation methods are used to enhance the biologically driven removal of toxic compounds from environmental matrices. The effectiveness of biodegradation is dependent on several factors, among them the characteristics of the soil, the bioavailability of the contaminants, and their chemical properties.

An important way of phenoxy herbicide removal from soil is by the use of indigenous soil bacteria harboring desirable catabolic genes. The first step in the phenoxy herbicide biodegradation pathway is initiated by α-ketoglutarate-dependent dioxygenase, an enzyme encoded by *tfdA* or *tfdA*-like genes [37] located in the *tfdABCDEF* gene cluster [38].

In recent decades, increasingly rapid advances in the application of molecular analysis in environmental studies have helped identify the bacterial communities involved in phenoxy herbicide biodegradation (**Table 4**). The bacteria able to metabolize phenoxy herbicides have been classified into three groups as follows: according to their physiology, employed degrading enzymes, and evolutionary origin [39–41] (**Table 4**).

1. The first group consists of fast-growing copiotrophic bacteria belonging to β- and γ-proteobacteria harboring the *tfdA* gene (e.g., *Cupriavidus necator* JMP134, *Burkholderia* sp. strain RASC, and *Rhodoferax* sp. strain P230). This first group has been subdivided into four subclasses according to *tfdA* sequence: *tfdA* Class I, II, and III [42, 43] and *tfdA* α [38]. Class I is found in *Cupriavidus pinatubonensis*; Class II is less widely distributed, being found only in *Burkholderia* spp.; and Class III is found in *Comamonas acidovorans* [38]. *TfdA*α was first identified in *Bradyrhizobium* sp. [40]. However, *tfdA*α-encoded protein has been described as α-ketoglutarate-dependent 2,4-D dioxygenase with lower activity than JMP134 dioxygenase.

2. The second group consists of slow-growing oligotrophic bacteria belonging to α-proteobacteria, phylogenetically closely related to *Bradyrhizobium* sp. [41], which were isolated from pristine environments. In this group, the phenoxy herbicide degradative gene was also identified and classified as *tfdA*α. Its gene sequence shows 50–60% similarity to the Group I degrader *Cupriavidus necator* JMP134.

3. The third group consists of bacteria belonging to the α-proteobacteria harboring the *tfdA*α gene, with *Sphingomonas* being the key member [41]. The wide diversity displayed by *tfdA*-like genes can partly be attributed to the wide range of bacteria (α-, β-, and γ-proteobacteria) capable of degrading phenoxy acids in the environment. Due to the high degree of homology between strains, the *tfdA* genes have been selected as biomarkers of the capability of bacteria to metabolize 2,4-D and MCPA [37, 38], and they are frequently used in studies of phenoxy acid biodegradation.

Much of the current literature on phenoxy herbicide metabolic pathways pays particular attention to the degradation pathway of 2,4-D. One of the most extensively studied 2,4-D degraders is *Cupriavidus necator* JMP134, known to harbor the 80-kb pJP4 plasmid. pJP4 carries all of the structural and regulatory genes needed to convert phenoxy herbicides to 2-chloromaleylacetic acid [56]. The *tfdA* fragment is responsible for the conversion of 2,4-D to 2,4-DCP [57]. Subsequently, 2,4-DCP is hydrolyzed to 3,5-dichlorocatechol by 2,4-DCP hydroxylase, which is encoded by *tfdB*. 3,5-Dichlorocatechol is further degraded via a pathway encoded by *tfdCDEF*.

Far too little attention has been paid to the metabolism of MCPA. MCPA degradation takes place by the cleavage of an ether linkage, resulting in the formation of the major metabolite, 4-chloro-2-methylphenol (MCP), and acetic acid [47]. This process is preceded by the expression of the *tfdA* gene. Mierzejewska et al. [34] report that microorganisms demonstrating the presence of *tfdA*α and *tfdA* Class III genes in soil contaminated with a commercial product containing MCPA displayed biodegradation potential.

The bacteria carrying *cad* genes, which encode the non-heme iron oxygenase, also have the potential to degrade both herbicides. The *cadRABKC* gene cluster was first identified and characterized in strain *Bradyrhizobium* sp. HW3 which was isolated from pristine environment in Volcanoes National Park, Hawaii [39]. So far, however, there has been little research on the mode of action and exact function of *cad* genes. According to Kitagawa et al. [38], *cadA*, *cadB*, and *cadC* genes are responsible for multicomponent oxygenase production, whereas *cadR* is a transcriptional regulator gene, which regulates the transcription of *cadABKC* in the presence of 2,4-D or 4-chlorophenoxyacetic acid. The *cadA* gene products show structural and functional differences to the *tfdA* gene with regard to their substrate preferences. Both the *cadA* and *cadB* and the *tfdA* genes code for aromatic ring

Class	Strain	Origin	Studied compounds	Identified functional genes	Source
α-Proteobacteria	*Sphingomonas paucimobilis*	Soil from Michigan (USA)	2,4-D	—	[44]
	Sphingomonas agrestis 58–1	Soil from Fukuoka Prefecture (Japan)	2,4-D, MCPA	*cadA*, *cadB*	[45]
	Bradyrhizobium sp.; *Sphingomonas* sp.	Root nodules; pristine environments (Hawaii, central California, USA; southwestern Australia, southwestern Africa; central Chile; northern Saskatchewan, Canada; northwestern Russia); volcanic soil (National Park (Kipuka Keana Bihopa, Hawaii, USA)	2,4-D	*tfdA*α, *cadA*, and *cadB*	[38, 39, 41]
	Sphingomonas sp.	Sediment from an aquifer in Fladerne Creek (Denmark)	MCPA	*cadA* and *cadB*	[46, 47]
β-Proteobacteria	*Comamonas acidovorans* strain MCI	Herbicide-contaminated building rubble (Germany)	2,4-D and MCPA	*tfdB* and *tfdC* genes	[48]
	Variovorax paradoxus	Soil from the Dijon INRA experimental station (France)	2,4-D	*tfdA*, *tfdB*, and *tfdR*	[49]
	Delftia sp.	Polluted river in Buenos Aires (Argentina)	2,4-D	—	[50]
	Cupriavidus campinensis BJ71	2,4-D-enriched soils from wheat fields in Beijing exposed for 2,4-D for at least 10 years (China)	2,4-D	Class I *tfdA* gene	[51]
	Achromobacter sp. LZ35	Soil in a disused pesticide factory in Suzhou (China)	2,4-D and MCPA	*tfdA* and *tfdB*	[52]
	Halomonadaceae sp.	Alkali Lake site in Oregon (USA) contaminated with 2,4-D production wastes	2,4-D	*tfdA*	[53]
γ-Proteobacteria	*Pseudomonas pickettii*	Agricultural soil from Michigan (USA)	2,4-D	—	[54]
	Pseudomonas maltophilia	Wheat rhizosphere (laboratory experiment)	2,4-D	—	[55]

Table 4.
Bacteria degrading phenoxy herbicides isolated from pristine and contaminated environments.

hydroxylation dioxygenases (RHDO), which are widely distributed in a number of microorganisms and might be transferred through horizontal gene transfer [38]; *CadA*- and *cadB*-encoded proteins are involved in the same initial step of 2,4-D degradation; however, the enzyme subunits have a different mode of action to the ketoglutarate-dependent dioxygenase encoded by *tfdA*. *CadA* and *cadB* were mostly identified in bacteria belonging to Groups I and II of phenoxy herbicide degraders. The products of *cadA* gene expression are able to initiate the degradation of both MCPA and 2,4-D. Furthermore, the abundance of *cadA* gene stimulates MCPA degradation [47]. The *cadA* gene is also essential for 2,4-D conversion in pure cultures of α-proteobacteria [38, 45, 58], and the *cadB* gene is also thought to play a sole role in the phenoxy acid degradation; however, the exact role of the *cad* genes remains not fully understood. The two genes share ~50% identity with *tfdA*, and it has been found that *cadA*, *cadB*, and *tfdA* are expressed simultaneously during MCPA degradation. Interestingly, some bacteria harbor all three *cadA*, *cadB*, and *tfdA* genes, thereby displaying a dual system of degradative genes [47].

The microbial degradation metabolic pathway of phenoxy herbicides has been elaborated in recent years (**Figure 3**). The first step of this catabolic pathway is initiated by either the *tfdA* gene which encodes α-ketoglutarate-dependent dioxygenase or *cadAB* genes which encode subunits of non-heme iron oxygenase [47]. Although these enzymes use different modes of action, both catabolic proteins have been shown to perform the same initial step in phenoxy acid degradation, turning 2,4-D into 2,4-DCP and MCPA into MCP.

Until recently, there has been little interest in the stereospecific Fe-(II) α-ketoglutarate-dependent dioxygenases which are encoded by *rdpA* and *sdpA* genes. These enzymes are described in literature as the ones which can also initiate the first step of the MCPA and 2,4-D degradation pathway. They were identified in *Delftia acidovorans*, *Rhodoferax* sp., and *Sphingobium* sp. Although the proteins encoded by the *rdpA* and *sdpA* genes possess the highly conserved amino acid sequence motif of *tfdA*-encoded proteins, they share only 37% identity with the *tfdA* genes of *C. necator* JMP134 [60, 61].

In addition to the soil bacteria, soil microfauna can also profoundly affect the biodegradation of organic contaminants. An important example of this relationship is the activity of earthworms, which move through the soil, causing better aeration and increasing soil moisture. Hence, insofar as their activity can influence the profile of the microorganism communities in the soil, they can indirectly enhance the process of phenoxy herbicide aerobic bacterial degradation [61].

3.2 Phytoremediation

A steadily developing strategy for the in situ treatment of contaminated soils is phytoremediation. It is a cost-effective and environmentally friendly strategy that uses plants to transform or mineralize xenobiotics to less toxic or environmentally neutral compounds [62]. Plants play a crucial role in the development of soil structure and stabilization of fundamental soil ecosystem functions such as water flow [63]. They produce also an array of catabolic enzymes, which operate to protect the host organisms and detoxify xenobiotic compounds [64]. Therefore, phytoremediation not only contributes to the detoxification of the environmental matrices but also has a positive influence on the functioning of the entire ecosystem.

The process of contaminant absorption by plants depends on several factors, including regional climate, soil type, and the nature of the pollutant [65]. The selection of an appropriate plant species and cultivar is critical for effective removal of a given contaminant from soil [66, 67]. This choice of phytoremediation candidate should

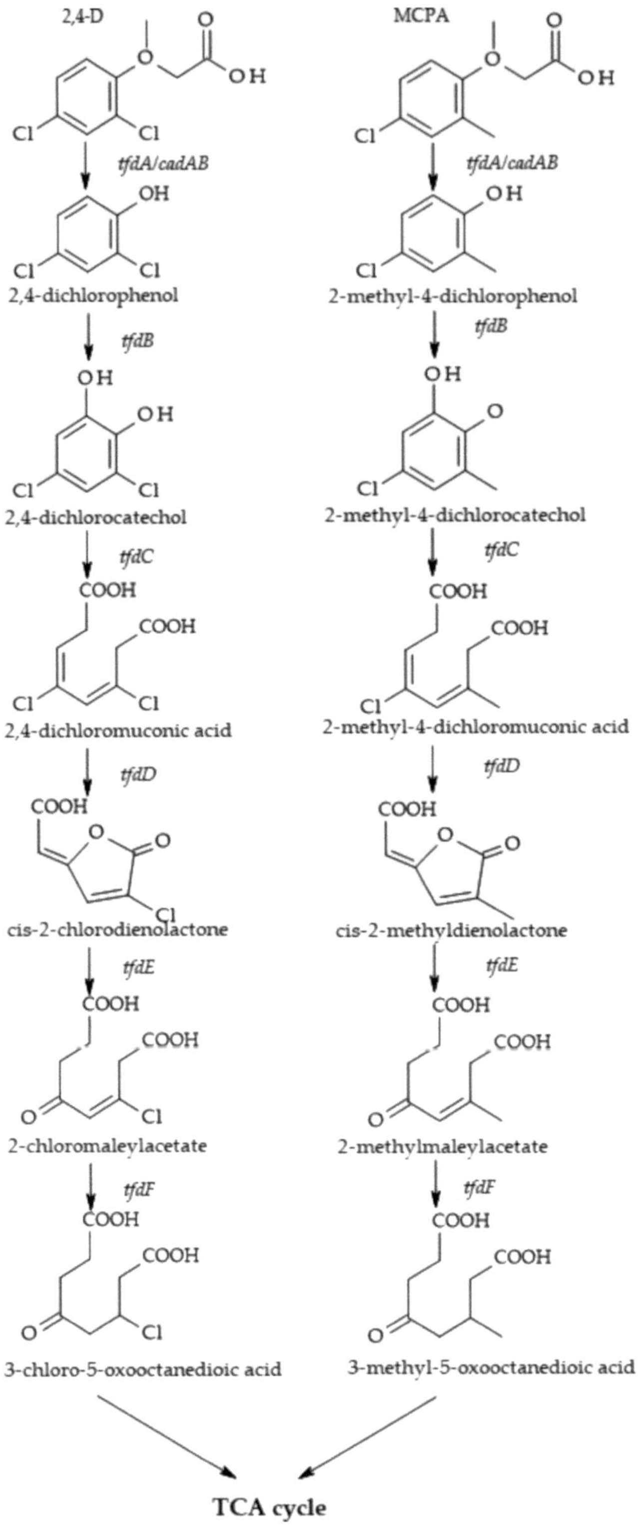

Figure 3.
Pathways of microbial degradation of 2,4-D and MCPA proposed by Pieper et al. [59]; in the picture there are indicated functional genes which encode catabolic enzymes as follows: tfdA, α-ketoglutarate-dependent dioxygenase; cadAB, subunits of non-heme iron oxygenase; tfdB, chlorophenol hydroxylase; tfdC, catechol 1,2-dioxygenase; tfdD, dichloromuconate cycloisomerase; tfdE, carboxymethylene butenolidase; tfdF, maleylacetate reductase.

particularly take into account plant growth rate, high biomass production, capacity for pollutant accumulation, and tolerance to higher xenobiotic concentrations [67].

In terms of phenoxy herbicide removal, there has been little investigation of the plant-mediated removal of 2,4-D and/or MCPA. For example, Ramborger et al. [68] evaluated the phytoremediation potential of *Plectranthus neochilus* (tea) exposed to the commercial pesticide containing 2,4-D (Aminol) in soil and water. The removal rate for 2,4-D reached 49% during 60 days, and the herbicide was not detected in plant leaves. Despite the fact that the phytoremediation potential of *P. neochilus* in soil was not sufficient, the plant exhibited satisfactory resistance to herbicide application. Moreover, the presence of phenolic compounds (e.g., ferulic and coumaric acid) in tea tissues indicated the ability of these plants to provide defense mechanisms against 2,4-D. The mechanism of the herbicide in plant begins by affecting the plasma membrane properties, subsequently leading to poor performance of mitochondria and peroxisomes [69]. In consequence, it stimulates the overexpression of abscisic acid (ABA) and ethylene biosynthesis genes, leading to significant changes of cellular redox potential by the production of reactive oxygen species (ROS) [70]. The occurrence of ROS leads to the production of phenolic compounds, i.e., ferulic acid and coumaric acid, which are responsible for the antioxidant self-defense mechanism of the plant against the herbicide. These phenolic compounds were found in higher concentrations only in plants that were exposed to 2,4-D and not in the controls.

3.3 Rhizoremediation

As mentioned above, plants play a key role in soil ecosystems by stabilizing the soil structure and by serving as primary sources of organic matter and energy which stimulate soil microbial activity [63]. Despite this, they are not the only contributors in the efficient phytoremediation of organic contaminants. Due to existing interactions between plant roots, root exudates, soil, and microorganisms, it has been proposed that the most effective method for the remediation of contaminated soil may be microbe-assisted phytoremediation (rhizoremediation).

Rhizoremediation is a naturally occurring process within the plant root zone (rhizosphere), where the growth of microorganisms and their degradative activity are stimulated by root exudates enriched by plant secondary metabolites (PSMs). Plant-derived compounds can [1] serve as primary substrates in cometabolism and provide energy for microbial growth [2], act as inducers of degradative enzymes due to their structural similarities to xenobiotics, and [3] enhance the degree of contamination removal by increasing pollutant bioavailability in soil [71].

The effectiveness of rhizospheral biodegradation depends also on the potential of the microorganisms inhabiting the rhizosphere to adapt to pollutant concentrations [72]. For effective degradation of contaminants to take place, a wide range of plants and bacterial traits is needed, involving the orchestrated interaction of a multitude of genes and enzymes. Rhizoremediation can therefore be optimized by selecting suitable plant-microbe sets, which can be achieved by combining plant and plant growth-promoting rhizobacteria (PGPR) [73] and/or microbes capable of contaminant degradation [74]. PGPR can improve phytoremediation efficiency by enhancing plant tolerance to various environmental stresses, promoting root growth and improving plant growth and health. In turn, some rhizospheral microorganisms can directly use their own degradative capabilities to metabolize organic pollutants [74, 75]. A study of rhizosphere-enhanced biodegradation of 2,4-D by Boyle et al. [76] found a significant difference in the mineralization of 2,4-D between monocot rhizosphere soils, dicot rhizosphere soils, and non-rhizosphere soils, with greater microbial activity being observed in monocot rhizosphere soil than in dicot rhizosphere soil or bulk soil. Therefore, both the soil and plant species determine the mineralization of tested contaminant. According to Shaw and Burns [77], the amendment of soil with 2,4-D increased the number of rhizospheric bacteria degrading 2,4-D in *Trifolium pratense* (red clover). Germaine et al. [78] also note the abundance of 2,4-D degraders in the stem and leaves of pea plant and that, under exposure to phenoxy herbicide, pea plants developed a stubby root system.

Furthermore, it has been hypothesized that PSMs may have a profound impact on the biodegradation of xenobiotics by providing the energy for microorganisms to carry out cometabolism; in this case, the xenobiotic is degraded as a secondary substrate [45, 71–73]. PSMs can be used as a primary source of carbon for bacterial communities to support their growth and stimulate the expression of desirable genes involved in the catabolic pathway of given xenobiotic. This is evident in the case of biphenyl, naringin, coumarin, myricetin, and l-carvone, which stimulate the activity of polychlorinated biphenyl (PCB)-degrading bacteria such as *A. eutrophus*, *Corynebacterium* sp., *P. putida* [79], and *Arthrobacter* sp. strain B1B [80]. Another example of PSM-stimulated PCB biodegradation was identified in mulberry (*Morus* sp.). In this case, the PSMs morusin, morusinol, and kuwanon C have been found to support the growth of the PCB-degrading bacterium *Burkholderia* sp. LB400 [81]. Likewise, the PSM (cumene) stimulates the activity of TCE-degrading *R. gordonia* bacteria [85]. According to Yi et al. [82], salicylic and linoleic acids, excreted by *Raphanus sativus*, enhanced the bioavailability of polycyclic aromatic hydrocarbons (PAHs) and increased the effectivity of their removal form soil. According to Ely and Smets [83], PAH biodegradation is stimulated by the presence of phenolic compounds, flavonoids, and gibberellic acid. Compounds such as acetophenone, phenethyl alcohol, p-hydroxybenzoic acid, and trans-cinnamic acid enhance the biotransformation of cis-1,2-dichloroethylene [84].

In addition, it has been hypothesized that PSMs may also induce the detoxification mechanisms taking place in bacterial cells [85, 86]. The expression of functional genes in bacteria is essential for the successful bioremediation of xenobiotics and can be stimulated by PSMs in different ways. However, very little information is given in the literature regarding the influence of PSMs on the induction of genes involved in catabolic pathways. Siciliano et al. [87] report greater induction of catabolic genes (*ndoB*, *alkB*, *xylE*) involved in the degradation of naphthalene in the rhizosphere soil of *Festuca arundinacea* (tall fescue) than in unplanted soil. Salicylate has been reported to have an upregulating effect on the expression of *bphA*, which encodes biphenyl dioxygenase in the PCB degrader *Pseudomonas* sp. Cam-1 [88]. The presence of salicylic acid was found to enhance the expression of the *bphA* gene in *R. eutropha* H850 and *P. fluorescens* P2W [89].

In addition, it has been hypothesized that the structural similarity between selected xenobiotics and PSMs may have a profound impact on the biodegradation of given, structurally related xenobiotic [71, 85]. For example, Urbaniak et al. [35] demonstrated the effect of a PSM, syringic acid, on the enhanced removal of structurally similar herbicide, MCPA, by indigenous soil bacteria, with greater MCPA depletion being achieved in samples enriched with PSM. The molecular analysis revealed ubiquitous enrichment of the samples with *Rhodoferax* spp., *Achromobacter* spp., *Burkholderia* spp., and *Cupriavidus* spp., which are commonly known as MCPA degraders. Also, a study by McLoughlin et al. [89] found the PSMs limonene and α-pinene to enhance 2,4-DCP degradation, but only following pre-exposure to both 2,4-DCP and monoterpene, with total 2,4-DCP mineralization extents of up to 71%.

Taking into account the abovementioned aspects, rhizoremediation can serve as a potential tool for phenoxy herbicide removal from soil ecosystems. However, to date, most studies have focused solely on the phyto- or biodegradation properties of plants or bacteria [71]. Consequently only limited data is available in terms of the impact of rhizoremediation on phenoxy herbicide removal from soil.

3.4 Endophyte-enhanced phytoremediation

Endophytic bacteria that reside inside plant tissues are also known to play a crucial role in the remediation of organic compounds. Plant-associated bacteria can enhance plant growth and degrade organic contaminants such as trichloroethylene and hydrocarbons [90]. The activity of endophytic bacteria can mitigate and improve plant conditions in stressful environments (such as contaminated soils). Field studies by Eevers et al. [91] showed that zucchini (*Cucurbita pepo*) plants inoculated with a consortium of three plant growth-promoting endophytic strains demonstrated an increased concentration of dichloro-bis(p-chlorophenyl)ethylene (DDE) in the aerial parts. The amount of DDE accumulated in *C. pepo* per growing season was significantly higher for inoculated plants. Thus such an approach might be promising for phytoremediation applications.

It has also been found that application of 2,4-D (1.42, 2.84, and 5.68 mg a.i./g soil) had a negative effect on the physio-morphological parameters of aerobic rice

Method	Compound	Initial concentration of compound used in an experiment	Duration of an experiment (days)	Removal of phenoxy-acetic acid (%)	Comments	Source
Bioremediation	2,4-D and MCPA	0.09 mmol/kg of soil	118	60	Activity of bulk soil microbial population from various soil samples	[93]
	2,4-D	1.8 kg/ha	10	45–48	Activity of bulk soil microbial population from clay and loamy soil samples	[94]
Phytoremediation and rhizoremediation	2,4-D	11.42 kg/ha	20	49	Use of *P. neochilus* for phytoremediation	[76]
	2,4-D 2,4-DCP	$1.22\ 10^{-3}$ µm $1.19\ 10^{-3}$ µmol	66	~60	Activity of rhizospheric soil bacteria derived from monocots	[68]
Endophyte-enhanced phytoremediation	2,4-D	47–360 mg/kg of soil	53	93–100	The inoculation of *P. sativum* by endophytic bacteria *P. putida* VM1450	

Table 5.
Biological remediation methods and % average removal of phenoxy herbicides from soil matrices.

(*Oryza sativa*) and reduced the number of plant endophytes [92]; however, inoculation of seeds with the endophytic bacteria strain *Stenotrophomonas maltophilia* improved plant characteristics under herbicide-stressed soils. *S. maltophilia* has previously been described as a plant growth-promoting endophytic strain with the ability to produce auxins and siderophores [92].

Bacterial endophyte-enhanced phytoremediation was also studied by Germaine et al. [78] on the example of *P. sativum*: plants were inoculated with genetically tagged endophytic bacteria, which naturally possess the ability to biodegrade 2,4-D. The inoculated plants not only displayed more efficient herbicide removal but also demonstrated a lack of 2,4-D accumulation in their aerial parts. Additionally the endophytic strain protected the pea plant from the toxic effects of 2,4-D, resulting in a greater increase of plant biomass and thus greater 2,4-D transportation to the aboveground parts of the plant from the soil.

Table 5 compares the presented biological methods of remediation of soils contaminated with phenoxy herbicides. It illustrates the differences of the removal of phenoxy herbicides from soil. It is apparent from this table that the most efficient method of contaminant removal is endophyte-assisted phytoremediation; however, more research on this topic needs to be undertaken before the association between role of symbiotic microorganisms and plants in removal of contaminants from environmental matrices is more clearly understood.

4. Conclusions

Uncontrolled use of phenoxy herbicides (2,4-D and MCPA) in the agricultural and gardening sector can result in their dispersal in soil and water ecosystems, which can significantly disturb the sustainability of the environment and increase its ecotoxicity level. Although their persistence in soil is limited due to their chemical characteristics, they can be transported and accumulated in water ecosystems through runoff and leaching. According to recent reports, phenoxy herbicides are especially toxic for plants, freshwater crustaceans, and amphibians; hence there is a growing need to limit the release of phenoxy acids in natural environments.

Taking into account the abovementioned aspects, the integration of bio-, phyto-, and rhizoremediation can serve as a potential tool for phenoxy herbicide removal from soil ecosystems. The ability of bacteria to metabolize phenoxy herbicides has been extensively studied over the last decades. However, to date, only limited data is available in terms of the impact of phyto- and rhizoremediation on phenoxy herbicide removal from soil. What is not yet clear is the impact of PSMs on the degradation of phenoxy herbicides. The similarity of the chemical structure of chosen PSMs and xenobiotics can be reflected in the xenobiotic degradation rates, e.g., the presence and induction of degradative genes and production of degradative enzymes, and the composition of microbial populations. To date, little evidence has been found associating the removal of phenoxy herbicides using both plants and microorganisms. However, the abovementioned research serves as a base for future studies on their application for the improvement of soil quality.

Considering the above, the chapter describes an interdisciplinary approach to tackling the problem of environmental phenoxy acid herbicide contamination through integrating available literature data on the physicochemical properties of 2,4-D and MCPA, as well as their levels in the environment and toxicity to the organisms from different trophic levels. It also outlines possible methods for their removal using nature-based techniques such as bio-, phyto-, and rhizoremediation.

Acknowledgements

This work was supported by the European Structural and Investment Funds, OP RDE-funded project "CHEMFELLS4UCTP" (No. CZ.02.2.69/0.0/0.0/17_050/0008485).

Conflict of interest

There is no conflict of interest.

Author details

Magdalena Urbaniak[1]* and Elżbieta Mierzejewska[2]

1 Department of Biochemistry and Microbiology, Faculty of Food and Biochemical Technology, University of Chemistry and Technology in Prague, Prague, Czech Republic

2 Department of Applied Ecology, Faculty of Biology and Environmental Protection, University of Lodz, Lodz, Poland

*Address all correspondence to: magdalena.urbaniak@vscht.cz

References

[1] Atwood D, Paisley-Jones C. 2008-2012 Market Estimates. Pestic Ind Sales Usage; 2017

[2] Eurostat. Sales of pesticides by type of pesticide

[3] Smith AE, Mortensen K, Aubin AJ, Molloy MM. Degradation of MCPA, 2,4-D, and other phenoxyalkanoic acid herbicides using an isolated soil bacterium. Journal of Agricultural and Food Chemistry. 1994;**42**(2):401-405

[4] Ma Q, Ding Y, Chang J, Sun X, Zhang L, Wei Q, et al. Comprehensive insights on how 2,4-dichlorophenoxyacetic acid retards senescence in post-harvest citrus fruits using transcriptomic and proteomic approaches. Journal of Experimental Botany. 2014;**65**(1):61-74

[5] Kamrin MA, editor. Phenoxy and benzoic acid herbicides. In: Pesticide Profiles. New York: CRC Press; 1997. pp. 299-224

[6] Skiba E, Wolf WM. Commercial phenoxyacetic herbicides control heavy metal uptake by wheat in a divergent way than pure active substances alone. Environmental Sciences Europe. 2017;**29**(1):26

[7] Paszko T, Muszyński P, Materska M, Bojanowska M, Kostecka M, Jackowska I. Adsorption and degradation of phenoxyalkanoic acid herbicides in soils: A review. Environmental Toxicology and Chemistry. 2016;**35**(2):271-286

[8] Gavrilescu M. Fate of pesticides in the environment and its bioremediation. Engineering in Life Sciences. 2005;**5**(6):497-526

[9] Parajulee A, Lei YD, Cao X, McLagan DS, Yeung LWY, Mitchell CPJ, et al. Comparing winter-time herbicide behavior and exports in urban, rural, and mixed-use watersheds. Environmental Science: Processes & Impacts. 2018;**20**(5):767-779

[10] Agency USEP. Reregistration Eligibility Decision (RED) 2,4-D. Washington, DC: EPA 738-R-05-002; 2005

[11] Gupta M, Garg NK, Joshi H, Sharma MP. Persistence and mobility of 2,4-D in unsaturated soil zone under winter wheat crop in sub-tropical region of India. Agriculture, Ecosystems and Environment. 2012;**146**(1):60-72

[12] Ignatowicz K, Struk-Sokołowska J. Sezonowe wahania zanieczyszczeń agrotechnicznych w rzece Narwi ze szczególnym uwzględnieniem herbicydów fenoksyoctowych. Środkowo-Pomorskie Tow Nauk Ochr Środowiska. 2004;**4**:189-205

[13] Birch GF, Drage DS, Thompson K, Eaglesham G, Mueller JF. Emerging contaminants (pharmaceuticals, personal care products, a food additive and pesticides) in waters of Sydney estuary, Australia. Marine Pollution Bulletin. 2015;**97**(1-2):56-66

[14] Ismail BS, Prayitno S, Tayeb MA. Contamination of rice field water with sulfonylurea and phenoxy herbicides in the Muda Irrigation Scheme, Kedah, Malaysia. Environmental Monitoring and Assessment. 2015;**187**(7):406

[15] Kucharski M, Domaradzki K. Changes in soil contamination by selected herbicides used in protection of cereals. Polish Journal of Soil Science. 2014;**47**(2):81-82

[16] Metcalfe CD, Helm P, Paterson G, Kaltenecker G, Murray C, Nowierski M, et al. Pesticides related to land use in watersheds of the Great Lakes basin. Science of the Total Environment. 2019;**648**:681-692

[17] Palma P, Matos C, Alvarenga P, Köck-Schulmeyer M, Simões I, Barceló D, et al. Ecological and ecotoxicological responses in the assessment of the ecological status of freshwater systems: A case-study of the temporary stream Brejo of Cagarrão (South of Portugal). Science of the Total Environment. 2018;**634**:394-406

[18] Chiron S, Comoretto L, Rinaldi E, Maurino V, Minero C, Vione D. Pesticide by-products in the Rhône delta (Southern France). The case of 4-chloro-2-methylphenol and of its nitroderivative. Chemosphere. 2009;**74**(4):599-604

[19] Kucharski M, Urbanowicz J. Badanie pozostałości linuronu i MCPA w glebie i roślinach ziemniaka. Biuletyn Instytutu Hodowli i Aklimatyzacji Roślin. 2008;**248**:61-66

[20] López-Roldán R, Jubany I, Martí V, González S, Cortina JL. Ecological screening indicators of stress and risk for the Llobregat river water. Journal of Hazardous Materials. 2013;**263**:239-247

[21] Salvo LM, Malucelli MIC, da Silva JRMC, Alberton GC, Silva De Assis HC. Toxicity assessment of 2,4-D and MCPA herbicides in primary culture of fish hepatic cells. Journal of Environmental Science and Health, Part B. Pesticides, Food Contaminants, and Agricultural Wastes. 2015;**50**(7):449-455

[22] Wei YD, Zheng HG, Hall JC. Role of auxinic herbicide-induced ethylene on hypocotyl elongation and root/hypocotyl radial expansion. Pest Management Science. 2000;**56**(5):377-387

[23] Pazmiño DM, Rodríguez-Serrano M, Romero-Puertas MC, Archilla-Ruiz A, del Río LA, Sandalio LM. Differential response of young and adult leaves to herbicide 2,4-dichlorophenoxyacetic acid in pea plants: Role of reactive oxygen species. Plant, Cell and Environment. 2011;**34**(11):1874-1889

[24] Martínez-Ruiz EB, Martínez-Jerónimo F. Exposure to the herbicide 2,4-D produces different toxic effects in two different phytoplankters: A green microalga (*Ankistrodesmus falcatus*) and a toxigenic cyanobacterium (*Microcystis aeruginosa*). Science of the Total Environment. 2018;**619**(620):1566-1578

[25] Li K, Wu JQ, Jiang LL, Shen LZ, Li JY, He ZH, et al. Developmental toxicity of 2,4-dichlorophenoxyacetic acid in zebrafish embryos. Chemosphere. 2017;**171**:40-48

[26] Sarikaya R, Yilmaz M. Investigation of acute toxicity and the effect of 2,4-D (2,4-dichlorophenoxyacetic acid) herbicide on the behavior of the common carp (*Cyprinus carpio* L., 1758; Pisces, Cyprinidae). Chemosphere. 2003;**52**(1):195-201

[27] Aronzon CM, Sandoval MT, Herkovits J, Pérez-Coll CS. Stage-dependent toxicity of 2,4-dichlorophenoxyacetic on the embryonic development of a south American toad, *Rhinella arenarum*. Environmental Toxicology. 2011;**26**(4):373-381

[28] Estevam EC, Nakano E, Kawano T, de Bragança Pereira CA, Amancio FF, de Albuquerque Melo AMM. Dominant lethal effects of 2,4-D in *Biomphalaria glabrata*. Mutation Research, Genetic Toxicology and Environmental Mutagenesis. 2006;**611**(1-2):83-88

[29] Bukowska B. Effects of 2,4-D and its metabolite 2,4-dichlorophenol on antioxidant enzymes and level of glutathione in human erythrocytes. Comparative Biochemistry and Physiology, Part C: Toxicology & Pharmacology. 2003;**135**(4):435-441

[30] Perkins EJ, Stiff CM, Lurquin PF. Use of *Alcaligenes eutrophus* as a

source of genes for 2,4-D resistance in plants. Weed Science. 1987;**35**(S1):12-18

[31] Taylor SG, Shilling DG, Quesenberry KH, Chaudhry GR. Phytotoxicity of 2,4-D and 2,4-dichlorophenol to red clover (*Trifolium pratense*). Weed Science. 1989;**37**(6):825-829

[32] Schweigert N, Hunziker R, Escher B, Eggen R. Acute toxicity of (chloro-) catechol-copper combinations in *Escherichia coli* corresponds to their membrane toxicity in vitro. Environmental Toxicology and Chemistry. 2001;**20**(2):239-247

[33] Podolska G. The effectiveness and phytotoxicity of herbicide in buckwheat cv. Kora. Polish Journal of Agronomy. 2014;**19**:17-24

[34] Mierzejewska E, Baran A, Urbaniak M. The influence of MCPA on soil phytotoxicity and the presence of genes involved in its biodegradation. Archives of Environmental Protection. 2017;**44**(4):58-64

[35] Urbaniak M, Mierzejewska E, Tankiewicz M. The stimulating role of syringic acid, a plant secondary metabolite, in the microbial degradation of structurally-related herbicide, MCPA. Peer J. 2019;**7**:e6745

[36] Weerakoon HPAT, Atapaththu KSS, Asanthi HB. Toxicity evaluation and environmental risk assessment of 2-methyl-4-chlorophenoxy acetic acid (MCPA) on non-target aquatic macrophyte *Hydrilla verticillata*. Environmental Science and Pollution Research. 2018;**25**(30):30463-30474

[37] Kitagawa W, Kamagata Y. Diversity of 2,4-dichlorophenoxyacetic acid (2,4-D)-degradative genes and degrading bacteria. In: Nojiri H, Fukuda M, Tsuda M, Kamagata Y, editors. Biodegradative Bacteria: How Bacteria Degrade, Survive, Adapt, and Evolve. Japan: Springer; 2014. pp. 43-57

[38] Kitagawa W, Takami S, Miyauchi K, Masai E, Kamagata Y, Tiedje JM, et al. Novel 2,4-dichlorophenoxyacetic acid degradation genes from oligotrophic bradyrhizobium sp. strain HW13 isolated from a pristine environment. Journal of Bacteriology. 2002;**184**(2):509-518

[39] Kamagata Y, Fulthorpe RR, Tamura K, Takami H, Forney LJ, Tiedje JM. Pristine environments harbor a new group of oligotrophic 2, 4-dichlorophenoxyacetic acid-degrading bacteria. Applied Environmental Microbiology. 1997;**63**(6):2266-2272

[40] Itoh K, Kanda R, Sumita Y, Kim H, Kamagata Y, Suyama K, et al. tfdA-like genes in 2,4-dichlorophenoxyacetic acid-degrading bacteria belonging to the *Bradyrhizobium-Agromonas-Nitrobacter-Afipia* cluster in α-proteobacteria. Applied and Environmental Microbiology. 2002;**68**(7):3449-3454

[41] Itoh K, Tashiro Y, Uobe K, Suyama K, Yamamoto H. Root nodule *Bradyrhizobium* spp. harbor acid-degrading proteins homologous with genes encoding 2, 4-dichlorophenoxyacetic acid-degrading proteins. Applied and Environmental Microbiology. 2004;**70**:2110-2118

[42] Mcgowan C, Fulthorpe R, Wright A, Tiedje JM. Evidence for interspecies gene transfer in the evolution of 2,4-dichlorophenoxyacetic acid degraders. Applied and Environmental Microbiology. 1998;**64**(10):4089-4092

[43] Poll C, Pagel H, Devers-Lamrani M, Martin-Laurent F, Ingwersen J, Streck T, et al. Regulation of bacterial and fungal MCPA degradation at the soil-litter interface. Soil Biology and Biochemistry. 2010;**42**(10):1879-1887

[44] Ka JO, Holben WE, Tiedje JM. Analysis of competition in soil among 2,4-dichlorophenoxyacetic acid-degrading bacteria. Applied and Environmental Microbiology. 1994;**60**(4):1121-1128

[45] Shimojo M, Kawakami M, Amada K. Analysis of genes encoding the 2,4-dichlorophenoxyacetic acid-degrading enzyme from *Sphingomonas agrestis* 58-1. Journal of Bioscience and Bioengineering. 2009;**108**(1):56-59

[46] Gözdereliler E, Boon N, Aamand J, De Roy K, Granitsiotis MS, Albrechtsen HJ, et al. Comparing metabolic functionalities, community structures, and dynamics of herbicide-degrading communities cultivated with different substrate concentrations. Applied and Environmental Microbiology. 2013;**79**(1):367-375

[47] Nielsen TK, Xu Z, Gözdereliler E, Aamand J, Hansen LH, Sørensen SR. Novel insight into the genetic context of the cadAB genes from a 4-chloro-2-methylphenoxyacetic acid-degrading *Sphingomonas*. Stevenson B, editor. PLoS ONE. 2013;**8**(12):e83346

[48] Müller RH, Jorks S, Kleinsteuber S, Babel W. *Comamonas acidovorans* strain MC1: A new isolate capable of degrading the chiral herbicides dichlorprop and mecoprop and the herbicides 2,4-D and MCPA. Microbiological Research. 1999;**154**(3):241-246

[49] Vallaeys T, Albino L, Soulas G, Wright AD, Weightman AJ. Isolation and characterization of a stable 2,4-dichlorophenoxyacetic acid degrading bacterium, *Variovorax paradoxus*, using chemostat culture. Biotechnology Letters. 1998;**20**(11):1073-1076

[50] González AJ, Gallego A, Gemini VL, Papalia M, Radice M, Gutkind G, et al. Degradation and detoxification of the herbicide 2,4-dichlorophenoxyacetic acid (2,4-D) by an indigenous *Delftia* sp. strain in batch and continuous systems. International Biodeterioration and Biodegradation. 2012;**66**(1):8-13

[51] Han L, Zhao D, Li C. Isolation and 2,4-D-degrading characteristics of *Cupriavidus campinensis* BJ71. Brazilian Journal of Microbiology. 2015;**46**(2):433-441

[52] Long ZX, Yan Z, Xin Z, Li YS. Biodegradation of the herbicide 2, 4-dichlorophenoxyacetic acid by a new isolated strain of *Achromobacter* sp. LZ35. Current Microbiology. 2017;**74**(2):193-202

[53] Maltseva O, Mcgowan C, Fulthorpet R, Oriel P. Degradation of 2,4-dichlorophenoxyacetic acid by haloalkaliphilic bacteria. Microbiology. 1 May 1996;**142**(5):1115-1122

[54] Ka JO, Holben WE, Tiedje JM. Use of gene probes to aid in recovery and identification of functionally dominant 2,4-dichlorophenoxyacetic acid-degrading populations in soil. Applied and Environmental Microbiology. 1994;**60**(4):1116-1120

[55] Lappin HM, Greaves MP, Slatert JH. Degradation of the herbicide mecoprop [2-(2-methyl-4-chlorophenoxy) propionic acid] by a synergistic microbial community. Applied and Environmental Microbiology. 1985;**49**(2):429-433

[56] Bælum J, Henriksen T, Christian H, Hansen B, Jacobsen CS. Degradation of 4-chloro-2-methylphenoxyacetic acid in top- and subsoil is quantitatively linked to the class III tfdA gene. Applied and Environmental Microbiology. 2006;**72**(2):1476-1486

[57] Fukumori F, Hausinger RP. *Alcaligenes eutrophus* JMP134 "2,4-dichlorophenoxyacetate monooxygenase" is an

alpha-ketoglutarate-dependent dioxygenase. Journal of Bacteriology. 1993;**175**(7):2083-2086

[58] Itoh K, Tashiro Y, Uobe K, Kamagata Y, Suyama K, Yamamoto H. Root nodule *bradyrhizobium* spp. Harbor tfdA and cadA, homologous with genes encoding 2,4-dichlorophenoxyacetic acid-degrading proteins. Applied and Environmental Microbiology. 2004;**70**(4):2110-2118

[59] Pieper DH, Reineke W, Engesser K-H, Knackmuss H-J. Metabolism of 2,4-dichlorophenoxyacetic acid, 4-chloro-2-methylphenoxyacetic acid and 2-methylphenoxyacetic acid by *Alcaligenes eutrophus* JMP 134. Archives of Microbiology. 1988;**150**(1):95-102

[60] Schleinitz KM, Kleinsteuber S, Vallaeys T, Babel W. Localization and characterization of two novel genes encoding stereospecific dioxygenases catalyzing 2(2,4-dichlorophenoxy) propionate cleavage in *Delftia acidovorans* MC1. Applied and Environmental Microbiology. 2004;**70**(9):5357-5365

[61] Liu Y, Liu S, Drake HL, Horn MA. Consumers of 4-chloro-2-methylphenoxyacetic acid from agricultural soil and drilosphere harbor cadA, r/sdpA, and tfdA-like gene encoding oxygenases. FEMS Microbiology Ecology. 2013;**86**:114-129

[62] Gerhardt KE, Huang X, Glick BR, Greenberg BM. Plant science phytoremediation and rhizoremediation of organic soil contaminants: Potential and challenges. Plant Science. 2009;**176**:20-30

[63] Machado F, Anderson C, Meenken E, Gillespie R, Peterson M, Harold M. The importance of plants to development and maintenance of soil structure, microbial communities and ecosystem functions. Soil and Tillage Research. 2018;**175**:139-149

[64] Singer AC, Crowley DE, Thompson IP. Secondary plant metabolites in phytoremediation and biotransformation. Trends in Biotechnology. 2003;**21**(3):123-130

[65] Reshma AC, Krishna RR. Plant species identification for phytoremediation of mixed contaminated soils. Journal of Hazardous, Toxic, and Radioactive Waste. 2012;**19**:218-229

[66] Siwek M. Biologiczne sposoby oczyszczania srodowiska—fitoremediacja. Wiadomości Botaniczne. 2008;**52**(1/2):23-28

[67] Posmyk K, Urbaniak M. Fitoremediacja jako alternatywna metoda oczyszczania środowiska. Aura. 2014;**7**:10-12

[68] Ramborger BP, Ortis Gularte CA, Rodrigues DT, Gayer MC, Sigal Carriço MR, Bianchini MC, et al. The phytoremediation potential of *Plectranthus neochilus* on 2,4-dichlorophenoxyacetic acid and the role of antioxidant capacity in herbicide tolerance. Chemosphere. 2017;**188**:231-240

[69] Rodríguez-Serrano M, Pazmiño DM, Sparkes I, Rochetti A, Hawes C, Romero-Puertas MC, et al. 2,4-Dichlorophenoxyacetic acid promotes S-nitrosylation and oxidation of actin affecting cytoskeleton and peroxisomal dynamics. Journal of Experimental Botany. 2014;**65**(17):4783-4793

[70] Grossmann K. Auxin herbicides: Current status of mechanism and mode of action. Pest Management Science. 2009;**66**(2):113-120

[71] Musilova L, Ridl J, Polivkova M, Macek T, Uhlik O. Effects of secondary plant metabolites on microbial populations: Changes in community structure and metabolic activity in

contaminated environments. Iriti M, editor. International Journal of Molecular Sciences. 2016;**17**(8):1205

[72] Lugtenberg BJ, Dekkers L, Bloemberg GV. Molecular determinants of rhizosphere colonization by pseudomonas. Annual Review of Phytopathology. 2001;**39**(1):461-490

[73] Ahemad M, Kibret M. Mechanisms and applications of plant growth promoting rhizobacteria: Current perspective. Journal of King Saud University—Science. 2014;**26**(1):1-20

[74] Kuiper I, Lagendijk EL, Bloemberg GV, Lugtenberg BJJ. Rhizoremediation: A beneficial plant-microbe interaction bioremediation: A natural method. Molecular Plant-Microbe Interactions. 2004;**17**(1):6-15

[75] Glick BR. Using soil bacteria to facilitate phytoremediation. Biotechnology Advances. 2010;**28**(3):367-374

[76] Boyle JJ, Shann JR. Biodegradation of phenol, 2,4-DCP, 2,4-D, and 2,4,5-T in field-collected rhizosphere and nonrhizosphere soils. Journal of Environmental Quality. 1995;**24**(4):782

[77] Shaw LJ, Burns RG. Enhanced mineralization of [U-14C]2,4-dichloropheeoxyacetic acid in soil from the rhizosphere of *Trifolium pratense*. Applied and Environmental Microbiology. 2004;**70**(8):4766-4774

[78] Germaine KJ, Liu X, Cabellos GG, Hogan JP, Ryan D, Dowling DN. Bacterial endophyte-enhanced phytoremediation of the organochlorine herbicide 2,4-dichlorophenoxyacetic acid. FEMS Microbiology Ecology. 2006;**57**(2):302-310

[79] Donnelly PK, Hegde RS, Fletcher JS. Growth of PCB-degrading bacteria on compounds from photosynthetic plants. Chemosphere. 1994;**28**(5):981-988

[80] Gilbert ES, Crowley DE. Plant compounds that induce polychlorinated biphenyl biodegradation by *Arthrobacter* sp. strain B1B. Applied and Environmental Microbiology. 1997;**63**(5):1933-1938

[81] Leigh MB, Fletcher JS, Fu X, Schmitz FJ. Root turnover: An important source of microbial substrates in rhizosphere remediation of recalcitrant contaminants. Environmental Science & Technology. 2002;**36**(7):1579-1583

[82] Yi H, Crowley DE. Biostimulation of PAH degradation with plants containing high concentrations of linoleic acid. Environmental Science & Technology. 2007;**41**(12):4382-4388

[83] Ely CS, Smets BF. Bacteria from wheat and cucurbit plant roots metabolize PAHs and aromatic root exudates: Implications for rhizodegradation. International Journal of Phytoremediation. 3 Oct 2017;**19**(10):877-883

[84] Fraraccio S, Strejcek M, Dolinova I, Macek T, Uhlik O. Secondary compound hypothesis revisited: Selected plant secondary metabolites promote bacterial degradation of cis-1,2-dichloroethylene (cDCE). Scientific Reports. 2017;**7**(1):8406

[85] Hu C, Zhang Y, Tang X, Luo W. PCB biodegradation and bphA1 gene expression induced by salicylic acid and biphenyl with *Pseudomonas fluorescence* P2W and *Ralstonia eutropha* H850. Polish Journal of Environmental Studies. 2014;**23**(5):1591-1598

[86] Uhlik O, Musilova L, Ridl J, Hroudova M, Vlcek C, Koubek J, et al. Plant secondary metabolite-induced shifts in bacterial community structure and degradative ability in contaminated

soil. Applied Microbiology and Biotechnology. 2013;**97**(20):9245-9256

[87] Siciliano SD, Germida JJ, Banks K, Greer CW, Lafayette W. Changes in microbial community composition and function during a polyaromatic hydrocarbon phytoremediation field trial. Applied and Environmental Microbiology. 2003;**69**(1):483-489

[88] Master ER, Mohn WW. Induction of bphA, encoding biphenyl dioxygenase, in two polychlorinated biphenyl-degrading bacteria, Psychrotolerant *Pseudomonas* strain Cam-1 and *Mesophilic Burkholderia* strain LB400. Applied and Environmental Microbiology. 2001;**67**(6):2669-2676

[89] McLoughlin E, Rhodes AH, Owen SM, Semple KT. Biogenic volatile organic compounds as a potential stimulator for organic contaminant degradation by soil microorganisms. Environmental Pollution. 2009;**157**(1):86-94

[90] Eevers N, Hawthorne JR, White JC, Vangronsveld J, Weyens N. Exposure of *Cucurbita pepo* to DDE-contamination alters the endophytic community: A cultivation dependent vs a cultivation independent approach. Environmental Pollution. 2016;**209**:147-154

[91] Eevers N, Hawthorne JR, White JC, Vangronsveld J, Weyens N, Hawthorne JR, et al. Endophyte-enhanced phytoremediation of DDE-contaminated using *Cucurbita pepo*: A field trial. International Journal of Phytoremediation. 21 Mar 2018;**20**(4):301-310

[92] Nahi A, Othman R, Omar D. Effects of Sb16 bacterial strain and herbicides on endophytic bacterial populations and growth of aerobic rice. Plant, Soil and Environment. 2016;**62**(10):453-459

[93] Bælum J, Prestat E, David MM, Strobel BW, Jacobsen CS, Park K. Modeling of phenoxy acid herbicide mineralization and growth of microbial degraders in 15 soils monitored by quantitative real-time PCR of the functional tfdA gene. Applied and Environmental Microbiology. 2012;**78**(15):5305-5312

[94] Boivin A, Amellal S, Schiavon M, van Genuchten MT. 2,4-Dichlorophenoxyacetic acid (2,4-D) sorption and degradation dynamics in three agricultural soils. Environmental Pollution. 2005;**138**(1):92-99

Chapter 6

Biodegradation of Sheep Wool Geotextiles Designed for Erosion Control

Jan Broda

Abstract

Wool geotextiles were formed from the meandrically arranged thick ropes and used as erosion control products. The geotextiles were installed in the experimental sites to protect the endangered slopes and the bank of ditches. Additionally, as a reinforcement of the soil, loose wool fibres were applied. The progress of wool biodegradation on the slope was investigated. Changes in the outer appearance, mechanical parameters, molecular structure and fibre morphology were analysed. Moreover, the nitrogen content in the soil and the effect of compounds released into soil on the grass growth were studied. The measurements revealed that the biodegradation starts at the cleavage of disulphide bonds, followed by disruption of the peptide bonds. Degradation is initiated in the outer cuticle and is followed by the decomposition of the inner cortical cells. During biodegradation, the nitrogen-rich compounds are released. The compounds act as an effective fertiliser which supports the growth of grass and significantly accelerates the greening of the slope.

Keywords: wool, geotextiles, erosion control, biodegradation, enzymes

1. Introduction

Sheep wool belongs to the oldest known natural fibres. The history of wool application coincides with the history of mankind. In many countries wool is highly valued, and manufacturing of wool products has a very long tradition.

For a long time, wool has been used for production of apparel textiles. Fine fibres have been used to manufacture high-quality fabric for luxury clothing. Coarser fibres were used to produce yarns used for knitting the traditional wool products. Coarser wool has been also successfully used for production of blankets, carpets and other interior textiles.

In the recent years, apart from the traditional products available on the market, new wool technical textiles appeared [1, 2]. Among them the most popular are acoustic and thermal materials used in the construction industry for insulation of pitched roofs, walls and ceilings [3–8]. For commercial application, also wool oil sorbents [9, 10], heavy metal-absorbing materials [11, 12] and geotextiles [13] are used.

Wool geotextiles include mats or blankets spread out on the ground and products which are buried in the soil. The mats are used to protect grass seeds in the ground. Due to wool's natural ability to regulate the temperature, mats create the proper microclimate for seed germination and, later, ensure favourable conditions for plant growth. The geotextiles buried in the soil are used in agriculture as plant fertilisers

IntechOpen

and materials to improve moisture retention [14]. Some products are applied as erosion control materials to protect unstable slopes and embankments [15, 16].

Wool geotextiles do not have a great global significance. The vast majority of geotextiles is produced from synthetic polymers and various biodegradable, natural materials. There are also products manufactured from the cellulosic fibres, mostly jute and coir [17].

In some cases, using wool for production of the geotextiles is reasonable and economically justified. As the product inseparably connected with sheep breeding, wool is available locally. Sometimes, due to low demand from the textile industry, the fibres are not further processed. There are attempts to use wool in agriculture as a nutrient promoting the plant growth [18–21] and organic nitrogen fertiliser [22], or it is mixed with clay to obtain environmentally friendly earthen construction materials [23–25]. However, these attempts are not always effective. Thus production of the wool geotextiles is a reasonable alternative, especially in the case of poor-quality wool treated as waste and a troublesome by-product of sheep breeding.

The geotextiles are exploited in contact with the soil in a humid environment. In such circumstances they are affected by the microorganisms naturally present in the soil. Microorganisms, both fungi and bacteria, secrete the extracellular enzymes which process the wool as a source of carbon, nitrogen and sulphur [26–28]. As a result, the wool is subject to gradual biodegradation. This process leads to the loss of mechanical strength and deterioration of the geotextile properties.

The biodegradation of geotextiles depends on the wool characteristic, product parameters and several environmental factors.

Sheep wool belongs to the group of the proteinaceous fibres which are built from the proteins formed by condensation of L-α-amino acids. A single fibre is composed from approximately 82% of keratinous proteins with high concentration of cystine and 17% of nonkeratinous proteins with a relatively low cystine content. Additionally, the wool contains approximately 1% by mass of non-proteinaceous material, mainly waxy lipids and polysaccharides [29].

A significant portion of the protein chains occur in the form of α-helix and form a compact structure of α-keratin, stabilised with different covalent and noncovalent bonds. The main bonds responsible for keratin stabilisation are disulphide bonds of cystine. The bonds form the intermolecular cross-links between different protein chains as well as intramolecular connections between different parts of the same protein chains. In addition to the disulphide bonds, the keratin structure is stabilised by the isopeptide linkages formed between the amino groups of lysine and carboxyl groups of aspartic or glutamic acid as well as by numerous noncovalent interactions: ionic bonds, hydrogen bonds and hydrophobic interactions.

The biodegradation of keratin is a two-stage process. In the first stage, the major part of the disulphide bonds, which prevent the protein-hydrolysing enzymes from accessing the peptide bonds, is broken. Then, keratin is hydrolysed by extracellular proteases, and the peptide bonds are disrupted. As a result, the soluble peptides which are further hydrolysed to particular amino acids are released. According to Kunert, the biodegradation caused by the fungus includes sulfitolysis and proteolysis [30, 31]. During sulfitolysis the disulphide bonds between polypeptide keratin chains cleave to S-sulfocysteine, and cysteine occurs. The reaction takes place under the alkaline conditions in the presence of inorganic sulphite produced by the fungus [32]. Sulfitolysis entails protein denaturation which facilitates the attack of keratinolytic proteases. As a result of protease activity, the denatured keratin proteins undergo proteolysis during which the peptide bonds are disrupted. Consequently, the simpler proteins, individual amino acids and ammonia, are released. According to Yamamura, in the first stage of degradation caused by the bacterium, the disulphide bonds get reduced by the disulphite reductase-like protein. The denatured keratin formed in the first stage is later decomposed by protease. Finally, the soluble products, peptides and/or amino acids, are released [33].

Wool fibres have complex morphology and consist of the outer layer of cuticle cells surrounding the inner cortex.

The cuticle cells consist of approximately 10% of the fibre weight. The flattened cells form the external layer and the characteristic scales on the fibre surface. The subsequent scales overlap in the longitudinal direction like tiles on a roof, with the edge pointing from the root to the tip of the fibre. The degree of overlap is about one-sixth of the length of the cuticle cells. For fine wool, except places where two cells overlap, the cuticle is normally one cell thick. In coarse fibres, the cuticle is thicker and consists of up to 15 layers of cells. Cuticle cells in fine wool range in thickness from 0.3 to 0.5 μm, while their length and width are about 30 μm and 20 μm, respectively [34].

The cuticle has a higher cystine content than the whole wool and is rich in cystic acid, serine, proline, glycine and valine [35]. The amino acids present in the cuticle belong to non-helix-forming amino acids which do not promote the formation of the α-helical structure. As a result, the cuticle structure is more amorphous than in the rest of the fibre.

The cuticle cells are composed of two distinct major layers: the outer exocuticle and the inner endocuticle. The exocuticle is the layer around 0.3 μm thick and represents approximately 60% of the total cuticle cell. The outer part of the exocuticle consists of a dense layer called the A layer. The exocuticle extends partly around the scale edges and contains the major amount of the cystine occurring in the cuticle. The endocuticle is around 0.2 μm thick and exhibits low cystine content, has a relatively low disulphide cross-link density and is classified as one of the nonkeratinous components of the fibre. The low cystine content makes the endocuticle more susceptible to chemical or enzymatic attack.

Individual cuticle cells are surrounded by a thin membrane called the epicuticle. The epicuticle is approximately 2–7 nm thick and is part of the resistant membrane system which surrounds all cuticle and cortical cells. The surface of the epicuticle is covered by a thin layer of a complex mixture of polar and non-polar lipids with long-chain fatty acids [36, 37]. The lipids are covalently bonded via the ester or thioester linkages and are responsible for the hydrophobic character of the surface of the wool fibres.

The cortical cells comprise the main bulk of almost 90% of the wool fibre. The long and spindle-shaped cortical cells are approximately 100 μm long and 3–6 μm wide. The cells are closely packed and oriented parallel to the fibre axis. The cells are composed of rod-like elements of crystalline proteins called intermediate filaments (microfibrils) embedded in an amorphous matrix. The intermediate filaments have the diameter of 7 nm and are built from the protofibrils formed by coiled coils of two or three α-helices of individual protein molecules. Simultaneously, the intermediate filaments are grouped together into the larger cylindrical units, the so-called macrofibrils with the diameter of about 0.3 μm [38].

The cuticle and the cortical cells are separated by the cell membrane complex (CMC). The CMC forms continuous network and provides adhesion between the particular cells. The membrane accounts for approximately 3.5% of the fibre, and its thickness is around 25 nm. The CMC is built from adhesive material that binds the cuticle and cortical cells together and consists of the central δ layer approximately 15 nm thick sandwiched by the two lipid layers called β-layers, each about 5 nm thick [39].

Due to various composition and cystine content, certain morphological elements exhibit various resistance to enzymes. Each part the biodegradation rate is different.

Based on the investigations on human hair which exhibits similar structure and morphology, various mechanisms for keratinous fibre degradation were proposed.

According to Kunert, degradation is initiated in the intercellular space in the cuticle. Degradation of the cell membrane complex is accompanied by the invasion of enzymes into the endocuticle. After the dissolution of the cell membrane complex, the individual cuticle cells are released and then gradually detached from one another. In the next step, the other parts of cuticle are slowly degraded. In the cortical layer, degradation starts in the intercellular spaces and progresses towards the keratin located between the macrofibrils. As a result, the separation of macrofibrils is observed. The separation is followed by the gradual degradation of macrofibrils, both from the surface and from the centre of the bundle. During the next stage, a gradual separation of individual microfibrils occurs [40, 41].

According to Wilson, degradation occurs in a predictable sequence, dependent on the composition and, hence, relative resistance of the hair structural elements to degradation by microorganisms. Degradation starts with the decomposition of cystine-poor components in both the cuticle and cortical cells. In the cuticle cells, the structural breakdown takes place in accordance to the sequence: cell membrane δ-layer, endocuticle, cell membrane β-layers, exocuticle, epicuticle and A layer. In the cortical cells, decomposition occurs as follows: cell membrane δ-layer, cell membrane α-layers, intermacrofibrillar matrix, microfibrils and intermicrofibrillar matrix [42].

According to Filipello Marchisio, biodegradation caused by fungi takes place in two possible mechanisms. The first mechanism, the so-called surface erosion, involves gradual destruction of the fibre from the external cuticle inwards to the cortical layer. According to this mechanism, the δ-layer of the cell membrane is attacked first. Then, the process continues by invading the hyphae under the scales. The hyphae lift up the cuticle and then secrete enzymes which digest the scales, starting from the cystine-poor endocuticle. In the next stage, the β-layers of cell membrane complex are digested, which is followed by destruction of the remaining cuticle layers. Then, the fungal hyphae attack the outer layers of the cortical cells. In this case, digestion appears to progress more rapidly. The cortex loses its compact structure, and the cells and macrofibrillar bundles of keratin become separated.

The second mechanism, called the radial penetration, involves random attack by variously specialised fungal hyphae which penetrate the fibre perpendicular to its surface. As a result, deep holes with a small diameter axis are formed [43, 44]. DeGaetano revealed that the holes exhibit minimal branching and their openings are placed under the scale edges or directly on the scale surface [45].

As for the product parameters, water absorption capacity is of a great importance for the course of wool biodegradation. For wool nonwovens, due to their porous structure with numerous open pores in the micron and submicron scales, water retention capacity reaches high values. Contrary to the woven and knitted fabrics, the porous structure geometry in the nonwovens is less uniform and is characterised by a greater distribution of pore size. The amount of absorbed water can be several times greater than the weight of the dry product. The water absorbed inside the nonwovens is stored for several days and then is slowly released into the environment. Persistent high humidity promotes the development of microorganisms, what significantly accelerates the biodegradation processes [46].

The most influential environmental factors include geographical location, ultraviolet exposure, temperature and humidity [47, 48]. Solazzo revealed that deterioration of textiles is dependent on the burial conditions, soil composition, pH, temperature, oxygen content and contact with wood coffins and metals [49].

Previous investigations on the biodegradation of wool in the soil focused on archaeological textiles [50, 51]. The literature lacks information on the biodegradation of the contemporary used wool geotextiles.

The chapter presents the results of investigations on the biodegradation of the wool geotextiles designed for erosion control. The geotextiles were installed in some experimental sites, and their biodegradation during 2 years of operation was studied.

2. Materials and methods

2.1 Materials

The locally available wool fibres with the length between 8 and 13 cm and the diameter between 25 and 32 μm sheared from the Polish mountain sheep were applied. Scoured fibres were used to produce a needle-punched nonwoven. The nonwoven was manufactured in industrial conditions by interlocking the wool carded web by means of the one-side needling machine operated from above with the punching density of 70 cm^{-2}. The product was the nonwoven with the thickness of 5.8 mm and the mass of 406 g/m^2.

The nonwoven strips were used for production of thick ropes with the diameter of 12 cm. The ropes were manufactured with the Kemafil machine [52]. The rope cores, made from densely packed wool nonwoven, were wrapped with a sheath made from cotton or thin polypropylene twine with the linear density of 240 tex.

The ropes were arranged in a meander-like pattern. Segments with the width of 2 m and the length adjusted to the length of the slope were prepared. In each segment, the subsequent turns of ropes were connected with five regularly spaced links. The polypropylene three-wire twine from fibrillated fibres with the linear density of 10 g/m was used for linking.

One slope was protected with the ropes made from a mixture of recycled fibres. The ropes installed on the slope were covered with the soil mixed with wool fibres. Mixtures of soil containing 0.25, 0.5 and 1% of wool in relation to dry soil weight were used.

2.2 Methods

During exploitation of the geotextiles, the behaviour of slopes was regularly monitored. The development of vegetation, density of the green cover as well as its colour were visually evaluated. Simultaneously, the external appearance of the ropes and the nonwovens was evaluated organoleptically. Periodically, the samples of geotextiles from randomly selected locations were taken for laboratory tests. Before examinations the samples were dried at room temperature and then mechanically cleaned from plant roots and soil particles. During the examinations the mechanical parameters of the nonwoven were determined. The tensile strength and elongation at break were measured in accordance with the Polish standard PN-EN ISO 10319:2010 [53]. The measurements were carried out along and across the nonwoven by means of the KS50 Hounsfield tensile machine equipped with the wide jaws. The static puncture resistance was determined by the California bearing ratio (CBR) test, in accordance with the Polish standard PN-EN ISO 12236:2006 [54]. The dynamic puncture resistance was measured using the cone drop test, in accordance with the Polish standard PN-EN ISO 13433:2006 [55].

The fibre morphology was studied by means of the scanning electron microscopy (SEM). A scanning electron microscope JEOL JSM 5500 LV was applied. The microscope was operated in the backscattered electron mode. Prior to observations, the fibres were sputtered with gold in JEOL JFC 1200 ionic sputter.

The chemical composition of wool keratin was analysed using the Fourier-transform infrared spectroscopy (FTIR). The FTIR spectrometer Nicolet 6700 equipped with a mirror beam collimator was applied. The measurements were carried out for tablets formed from the chopped fibres blended with powdered sodium chlorine NaCl. The spectra were registered in the range from 400 to 4000 cm^{-1}. During the analysis, the spectra were smoothed with the OMNIC software and normalised against the peak intensity. The band at 1451 cm^{-1} corresponding to CH_2 group was chosen for normalisation.

During the investigations, the nitrogen content in the soil was measured. The content of organic, ammonium (NH^{4+}), nitrite (NO^{2-}), nitrate (NO^{3-}) as well as Kjeldahl and total nitrogen was determined. The measurements were performed according to the Polish standards: PN-ISO 7150-1:2002, PN-73C-04576:2006, PN-82C-04576:1982 and PN-EN 25663:2001 [56–59].

2.3 Installation sites

The wool fibres and wool geotextiles were installed in four experimental sites situated in Bielsko-Biala and surroundings. All sites were located in southern Poland, the temperate climate zone.

In the preliminary research, the samples of ropes made from the wool nonwoven were buried in the experimental site located in the university campus. Then, the geotextiles formed from the meandrically arranged thick ropes were used for the renovation of the bank of the drainage ditch damaged by water surface erosion. In further investigations, the similar wool geotextiles were used for stabilisation of steep slope prone to sliding in the abandoned gravel pit. In both places, the segments of geotextiles were spread on the surface of the slopes and anchored to the slopes with metal "U-shape" pins. Finally, the geotextiles were covered with a layer of topsoil (**Figure 1**).

In recent examinations, geotextiles made from the recycled fibres were applied for protection of a steep slope exposed to intensive rill erosion. The slope was

a b

Figure 1.
Installation of the geotextiles, (a) bank of the drainage ditch in Miedzyrzecze, (b) slope in the abandoned gravel pit in Nieboczowy.

a b

Figure 2.
Installation of the geotextiles in Lipnik, (a) mixing fibres with the soil, (b) the geotextiles covered with a layer of soil mixed with fibres.

Site	Bielsko-Biala (B-B)	Miedzyrzecze (M)	Nieboczowy (N)	Lipnik (L)
Product form	Ropes	Ropes	Ropes	Fibres
Object	Experimental flat plot	Bank of the drainage ditch damaged by erosion	Unstable slope in the abandoned gravel pit	Slope between terraces damaged by erosion
Inclination	Flat	1:1.5	1:1.8	1:1.5
Length	—	6 m	5 m	5.5 m
Installation time	October 2014	May 2015	February 2016	April 2017
Covering soil	Sandy silty clay	Clay of low plasticity	Clay of high plasticity	Clay of low plasticity
pH	7.2	7.1	7.0	7.0
Seeding	Yes	No	No	Yes

Table 1.
Experimental sites and their characteristics

located between flat terraces artificially formed on a gently sloping hill. In this case, the geotextiles were covered with soil reinforced with loose wool fibres (**Figure 2**).

In two locations (Bielsko-Biala and Lipnik), the grass seeds were sown on the installed geotextiles. The seed was a commercial mixture of perennial ryegrass (*Lolium perenne*), commonly used for slope stabilisation.

Basic characteristic of the experimental sites is presented in **Table 1**. More details concerning the site characteristic and installation procedure were presented in the previous publications [60–65].

3. Results

3.1 Slope greening

In site B-B, the samples of ropes were buried in the ground in autumn. In the spring, with the beginning of the vegetation season, the growth of grass was initiated. After a few weeks, the plot was covered with green. In places where wool ropes were buried, the grass was higher and dark green (**Figure 3a**).

In site M, the geotextiles were installed in late spring. In the first year, during the whole vegetation season, rare and miserable self-sown plants appeared on the bank. In the following vegetation season, the ditch bank was covered with high and dense cover consisting of a mixture of grasses and various local herbaceous plants. The plants had an intense dark green colour, much darker compared to the plants grown in other parts of the bank (**Figure 3b**).

In site N, the rope segments were installed at the end of the winter season. In the early spring, first sparse seedlings of local species appeared on the slope. Then, in the following summer weeks, the slope was covered with a dense green cover of grasses and other local herbaceous plants (**Figure 3c**). Next year the slope became green and covered with lush vegetation. The plants reached the high altitude and formed a very dense, uniform cover on the whole surface of the slope. In comparison to the previous year, the vegetation cover was much denser. Apart from the species observed in the previous season, more than 20 other species were identified.

In site L, the geotextiles made from recycled fibres were covered with the soil mixed with wool in the early spring. Shortly after sowing, the quick growth of grass was observed. About 2 months later, the slope was completely greened and

Figure 3.
Influence of wool on slope greening, (a) Bielsko-Biala (6th month), (b) Miedzyrzecze (12th month), (c) Nieboczowy (4th month), (d) Lipnik (3rd month).

	Nitrogen content [mgN/g]					
	Organic	Ammonium NH_4^+	Nitrite NO_2^-	Nitrate NO_3^-	Kjeldahl	Total
1	615	25.2	1.2	0	640	641
2	2481	29.1	1.4	0	2510	2511
3	1396	33.7	1.6	207	1430	1639
4	1338	105.5	2.7	84	1444	1530

1, soil without fibres; 2, soil mixed with wool; 3, soil mixed with wool after 6 months; 4, soil mixed with wool after 12 months

Table 2.
Nitrogen content in the soil mixed with wool used for covering of ropes on the slope in Lipnik

coated with a dense grass cover. The highest and the most dense grass cover with an intense dark green colour appeared in the places protected with the soil mixed with wool (**Figure 3d**).

The ropes were covered with native soil poor in nitrogen compounds. The total nitrogen content in the soil before installation on the slope was low and included mainly nitrogen in an organic form (**Table 2**). Once the wool was added, the nitrogen content in the soil, mainly organic nitrogen, increased by fourfold. After 6 months of using the slope, the organic nitrogen content in the ground decreased by almost a half. At the same time, a slight increase in the content of ammonium nitrogen and a significant increase in nitrate nitrogen were observed. In the second year of exploitation, after 1 year the total nitrogen content in the soil mixed with wool was slightly lower. In this time, the majority of nitrogen occurred in the organic form. The soil also contained a significant amount of ammonium and nitrate form.

3.2 External appearance

In site B-B, after first 12 months of exploitation, the nonwoven did not show external signs of damage. In this time the ropes made from the nonwoven kept their physical permanence. Then, in the following months, the discolorations of wool appeared. The fibres forming the nonwoven became fragile, and the nonwoven could easily be torn in hands.

In site M, the first visible signs of degradation were observed after 1 year of exploitation. Serious damage of the nonwoven was revealed in many places on the bank. Plenty of gnawing holes, discolorations and numerous signs of rotting were visible in the nonwoven. In some places only the rotten remnants of the nonwoven were found. After 18 months the nonwoven completely lost its mechanical integrity.

In site N, there were no visible signs of damage of the nonwoven during the first year. The ropes maintained their shape and mechanical integrity. After 18 months, under the weight of the covering soil, the previously round ropes became flat. Simultaneously, in the nonwoven forming the outer layer of the ropes, small gnawing holes appeared.

In site L, shortly after mixing wool with the soil, the fibres became fragile. Few weeks after introducing into the soil, the destruction of fibres was already strongly advanced. After 6 months only short and broken remnants of the fibres were visible.

3.3 Mechanical parameters

Table 3 presents basic mechanical parameters: tenacity and elongation at break of wool nonwovens before installation and after several months of exploitation in the ground.

For the ropes buried in the site B-B, the tenacity of the nonwoven decreased by 51% after 6 months. After 12 months the reduction of the nonwoven tenacity reached 95%. At the beginning the elongation did not change, and then, after 12 months, it decreased by 58%. For the ropes buried in the ground for a longer time, the nonwoven was easily torn in hands, so it was impossible to measure its strength.

For the geotextiles installed in the site M, during first 6 months, the tenacity decreased by 28%. In the months that followed, the reduction of the nonwoven

Site	Months	Tenacity [kN/m]	Elongation at break [%]
—	0	0.67 ± 0.1	40 ± 3
Bielsko-Biała	6	0.34 ± 0.01	39 ± 3
	12	0.04 ± 0.01	17 ± 3
Miedzyrzecze	6	0.48 ± 0.02	28 ± 2
	12	0.05 ± 0.01	24 ± 2
Nieboczowy	6	0.62 ± 0.01	33 ± 3
	12	0.40 ± 0.01	31 ± 3
	18	0.07 ± 0.01	39 ± 3

Table 3.
Mechanical parameters of the wool nonwoven exploited on slopes

Site	Months	Puncture resistance	
		Static [kN]	Dynamic [mm]
—	0	0.16 ± 0.01	32 ± 0.5
Bielsko-Biała	6	0.02 ± 0.01	30 ± 0.5
	12	—	—
Miedzyrzecze	6	0.04 ± 0.01	38 ± 1
	12	—	—
Nieboczowy	6	0.09 ± 0.01	36 ± 3
	12	0.06 ± 0.01	38 ± 1
	18	0.07 ± 0.01	39 ± 1

Table 4.
Puncture resistance of wool nonwoven after exploitation on the slopes

tenacity was much higher and reached 93% after 1 year of exploitation. The reduction of the nonwoven tenacity was connected with the decrease of the elongation, 30 and 40% after 6 and 12 months, respectively. Similarly to the samples taken in site B-B, the nonwoven exploited for a longer time disintegrated easily in hands, and it was impossible to measure its strength.

For the samples exploited in the site N, the reduction of the nonwoven tenacity was relatively small in the first period and after first 6 months equalled to only 7%. Another 6 months later, it exceeded 35%. During the months that followed, a rapid drop of the nonwoven strength was detected. After 18 months the tenacity reached the minimal value which was 85% lower than the value measured for nonwoven taken after 1 year. The reduction of the nonwoven tenacity was connected with the increase of the elongation at break. The extreme increase of the elongation was observed for the nonwovens exploited in the soil for 18 months.

Table 4 presents the values of static and dynamic puncture resistance measured for the nonwovens before installation and after several months of exploitation in the ground.

Puncture resistance illustrate the resistance of the nonwovens to multidirectional loads. The measurement of static resistance simulates the pressure exerted by roots or stones onto the geotextiles lying on the ground. The test involves pushing a plunger through a flat specimen, while the penetration force is measured. The dynamic resistance measurements imitate dropping sharp stones onto the geotextile surface. The test involves dropping of steel cones from a fixed height while the hole diameter is measured.

The static puncture resistance of the nonwovens in two sites, B-B and M, reduced drastically by 88 and 75% already after 6 months of exploitation. In the same time, the dynamic puncture resistance for the specimen taken in B-B remained at the same level and for the specimen in M—increased by 19%. During further exploitation the nonwoven was weakened, and after 12 months of measuring, the puncture resistance was impossible.

As for the nonwoven exploited in the site N, the reduction of the puncture resistance occurred gradually over time and much less rapidly than in the previous sites. After 6 months, the penetration force—as the measure of the static puncture resistance—dropped by 44%. Then, after 12 months, the force decreased by 63% and remained at the same level even after 18 months of exploitation. Similarly, the dynamic puncture resistance changed during the first 6 months by 13%. After 12 months, the measure of the dynamic puncture resistance changed by 19% and remained at this level until 18 months passed.

3.4 Fibre morphology

On the surface of the raw wool fibres, the characteristic scales with smooth surface and sharp edges were visible. In the samples taken in the site B-B, irregular cracks appeared in many places on the surface of scales already after 1 month of exposure in the soil (**Figure 4a**).

Simultaneously, the erosion of the scale edges began. As the destruction progressed, after 6 months the scales became barely visible. In the next months, the outer cuticle layer was completely destroyed. Then, the particular macrofibrils with the diameter of few microns, parallel to the fibres axis, became visible. After 9 months, the macrofibrils still adhered tightly to one another. Later, after 12 months, the macrofibrils were more separated. After longer period of soil exposition, the microphotographs showed microorganisms digesting the wool structure. The microorganisms possessed relatively big dimensions and adhered locally to the fibre surface (**Figure 4b**).

In the fibre samples extracted in the site M after 6 months, the outer cuticle layer was heavily damaged (**Figure 5a**). In many fibres the scale edges were completely eroded, and their surface was partially removed. On some fibres transverse fractures or longitudinal cracks were visible. As a result of the mechanical damage, the outer cuticle was torn in some fibres. The cracks in the cuticle enabled quick enzymatic penetration, what led to the rapid fibrillization of the fibres. Surface of the fibres taken after 12 months was colonised by numerous fine microorganisms. The microorganisms

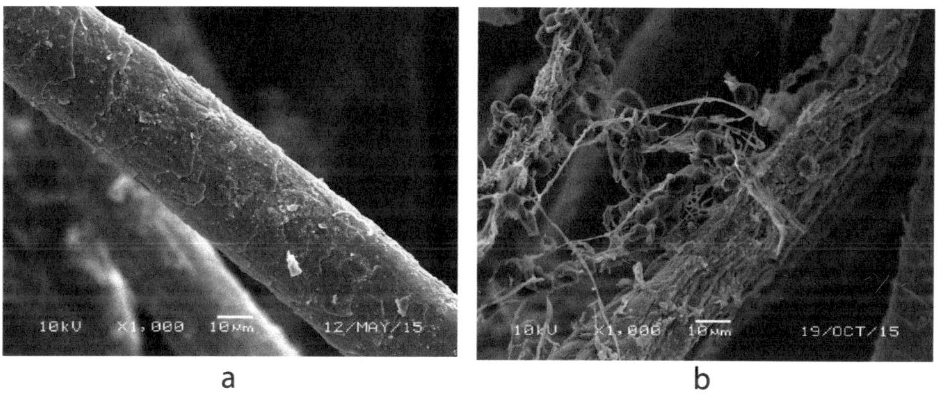

Figure 4.
Morphology of wool taken in Bielsko-Biala after (a) 6 months and (b) 12 months.

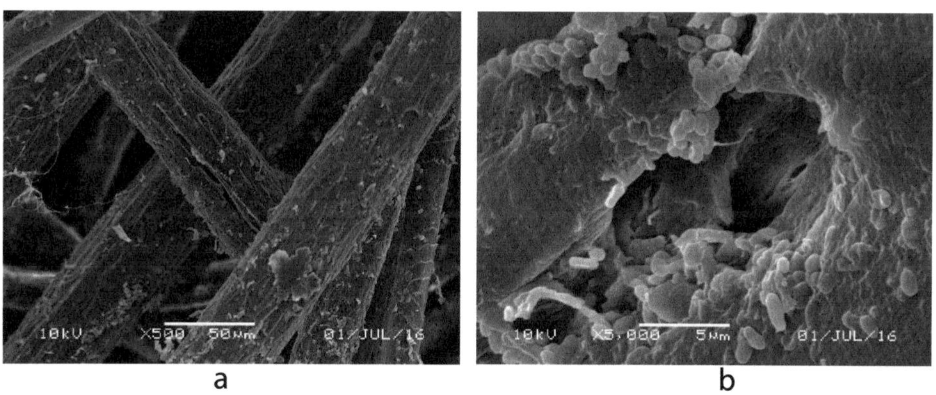

Figure 5.
Morphology of wool taken in Miedzyrzecze after (a) 6 months and (b) 12 months.

formed large colonies which covered almost the whole fibre surface. In fibres extracted after 12 months, the outer cuticle layer was completely destroyed. The scales were no longer visible, and, in many places, the inner fibrillar structure of the core was exposed. The fibres had many deep and widespread gnawed cavities (**Figure 5b**).

Unlike the previous sites, the wool taken from the site N after 6 months still has well visible scales. The significant changes in fibre morphology were observed after 1 year. The scales of all the fibres were barely visible, and the continuous protective thin layer covering the scales was completely destroyed. After 1 year some fibres were colonised with big colonies of microorganisms. The microorganisms formed a thick layer on many fibres, which occupied almost the whole fibre surface (**Figure 6a**).

In the fibres inhabited by microorganisms, large cavities penetrating deep inside the fibres were formed. About 18 months after the installation, the scales in almost all the fibres were completely removed, and the inner parts of the fibres were exposed. In many fibres, loosely connected separated fibrils were revealed (**Figure 6b**).

The fibres in the site L showed numerous signs of mechanical and biological damage already after 2 months. Mechanically damaged fibres were cracked, broken or frayed. The scales of many fibres were greatly damaged or even completely destroyed (**Figure 7a**). The deeper layered fragments forming the cortex layer were visible. After 6 months the majority of the fibres showed an advanced stage of fibrillization (**Figure 7b**). Large fragments of many fibres were missing. After 12 months the fibres were totally destroyed, and only fragmentary remnants of fibres could be found.

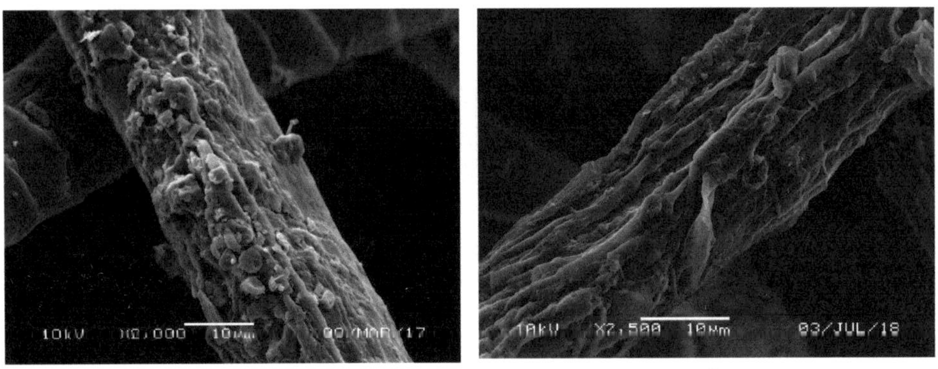

a b

Figure 6.
Samples of wool geotextiles taken in Nieboczowy after (a) 12 months and (b) 24 months.

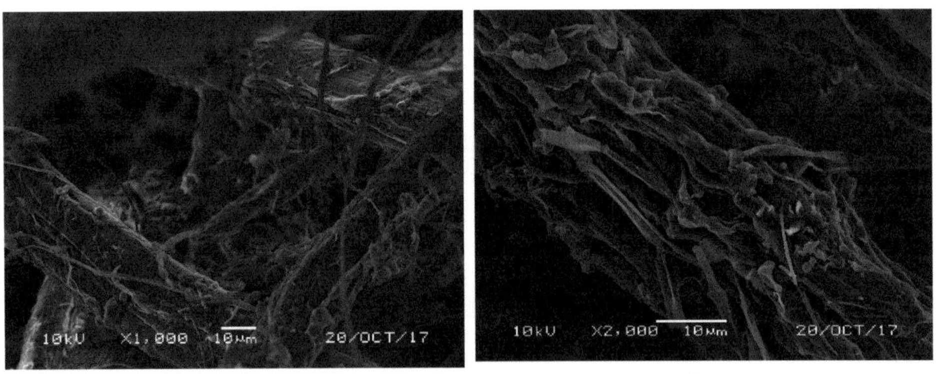

a b

Figure 7.
Samples of wool geotextiles taken in Lipnik after (a) 2 months and (b) 6 months.

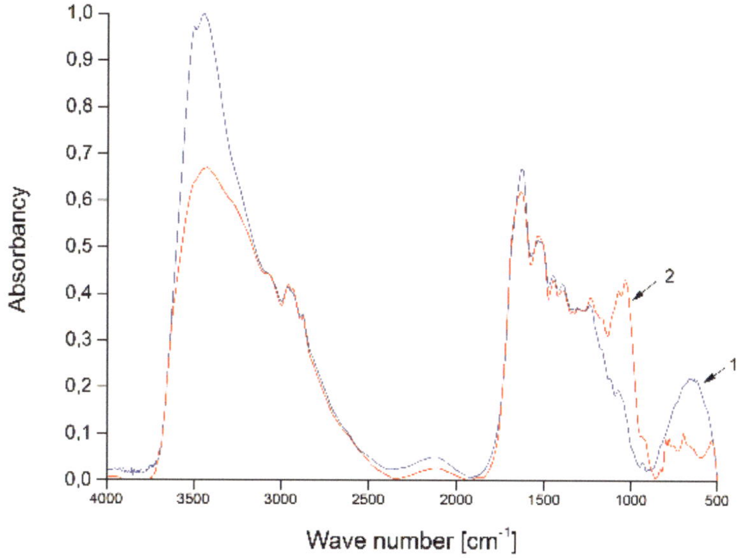

Figure 8.
FTIR spectra of wool used for mixing with soil to protect the slope in Lipnik; (1) raw fibres; (2) after 6 months.

3.5 Chemical structure

The FTIR spectra of all the investigated fibres showed typical bands assigned to the peptide bonds in the amide region. For the raw fibres, before installation of the geotextiles in the soil, strong amide I, II and III bands occurred at 1657 cm^{-1}, 1556 cm^{-1} and 1244 cm^{-1}, respectively. In the sulphoxide region at 1073 cm^{-1}, the band corresponding to S—O bond was observed. For the raw wool, the band possessed low intensity and was poorly visible.

The position and intensity of the amide bands did not change during the first 6 months of exploitation on slopes in three sites: B-B, M and N. The gradual reduction in the amide band intensity was noticed after 12 months. The specimens taken from the site N recorded a lower reduction of the band intensity. Simultaneously, with the decrease of the amide band intensity after 6 and 12 months, the increase of the intensity of the S—O peak was observed.

The fibres mixed with soil in the site L showed no change of the amide band intensity during the first 6 months. At the same time, the intensity of the S—O peak was considerably greater (**Figure 8**).

4. Discussion

Wool biodegradation was caused by the enzymes secreted by the microorganisms in the soil. On the molecular level, the enzymes first attacked the disulphide bonds of cystine. At the beginning, few weeks after installation of the geotextiles in the soil, disruption of the disulphide bonds occurred mostly in outer cuticle cells. Initially, the sharp edges of the scales eroded, and numerous cracks appeared on the hitherto smooth-scale surface. The complete destruction of the cuticle cells was observed in the next stage. The outer appearance of the geotextiles remained virtually unchanged; however, a significant reduction of the mechanical parameters was recorded.

Due to high cystine content, the biodegradation of cuticle runs relatively slowly. The outer cuticle of the geotextiles used in two sites, B-B and M, was destroyed

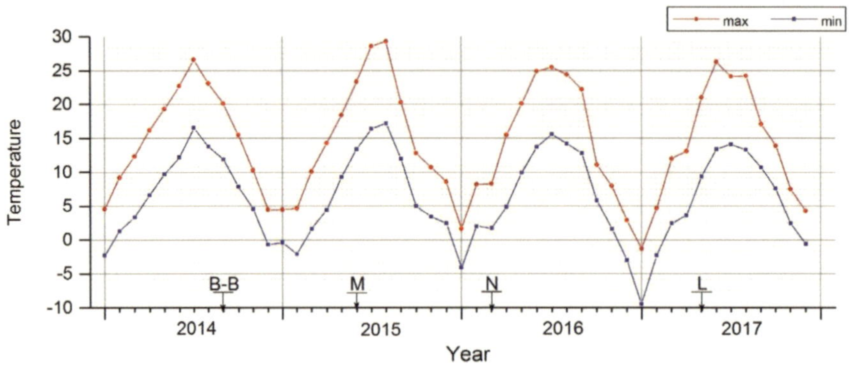

Figure 9.
Minimal and maximal temperature during operation of the slopes.

within 6 months. In B-B this period was in the winter with temperatures between −2 and 5°C (**Figure 9**). In M it was the summer during the vegetation season with temperatures between 10 and 30°C. As for site N, the time of scale destruction was extended to 12 months. This period included the vegetation season with temperatures between 10 and 25°C and winter with temperatures between 0 and −10°C.

Destruction of the cuticle occurred faster in fibres mixed directly with the soil—already 2 months after installation on the slope. In this case, fast destruction resulted from mechanical damage of the fibre scales, which occurred during mixing the fibres with the soil. This rapid degradation was also favoured by weather conditions, positive temperatures and high soil moisture ensured by the heavy rainfall.

During longer operation in the soil, the fibre surface was colonised by microorganisms. The microorganisms adhered to the fibres surface and secreted enzymes which—after disrupting the significant amount of disulphide bonds and denaturing the keratin structure—initiated the breakdown of the peptide bonds. After destroying the cuticle cells, the enzymes penetrated into the deeper layers of the fibres and caused gradual decomposition of the cortical cells. The process started with the breakdown of the amorphous proteins filling the spaces between fibrils. As a result, the fibrillar structure of the cortical cells was revealed. In the first stage, the fibrils adhered to each other. Over time, single fibrils became more separated. Simultaneously, deep and wide cavities appeared in many fibres as the results of enzyme activity.

At this stage, the changes recorded in fibre morphology were connected with geotextile discoloration, changes in the visual appearance and touch and a drastic drop of the mechanical strength.

The biodegradation of cortical cells occurred with a relatively higher velocity. Keratin in the cortical cells possesses lower content of cystine and exhibits smaller resistance to enzymatic attack. As a result, the separation of fibrils and further disintegration of keratin structure occurred relatively quickly—within 3 or 4 months.

Keratin destruction led to the release of the soil nitrogen compounds. Immediately after adding the fibres, the content of organic nitrogen in the soil increased four times. Then, during the exploitation of the slope, the organic nitrogen was transformed into mineral nitrogen forms absorbable by plants. Nitrogen compounds, which drive the growth of plants, considerably accelerated the greening of the slopes. Due to rapid wool biodegradation in site L, the influence of wool on intense grass growth was immediately observed 3 months after installation. In other places the effect was recorded later, in the new vegetation season.

5. Conclusions

The wool geotextiles used for slope and ditch protection remained in humid environment and were exposed to the enzymes secreted by the microorganisms naturally inhabiting the soil. Enzyme activity resulted in the gradual degradation of wool. The process was initiated in the cuticle cells forming the outer layer of the fibres. Due to a relatively high enzymatic resistance of proteins forming the cuticle, the destruction was carried out relatively slowly. It took at least 6 months to destroy the fibres which were not mechanically damaged. Until that time, the geotextiles lost their mechanical properties only to a minimum, maintaining their protective potential.

After a longer time and destruction of the outer cuticle, the enzymes penetrated the inner parts of the fibres. In a relatively short period, the separation of fibrils followed by the total disintegration of the fibre structure was observed. The changes observed in the fibre morphology were connected with a drastic reduction of the nonwoven mechanical strength, what caused the geotextiles losing their protective abilities.

At the advanced stage of the biodegradation, the enzymes catalysed the breakdown of wool keratin into the soluble peptides and certain amino acids. The nitrogen-rich organic compounds released into the soil were gradually transformed into mineral nitrogen forms absorbable by plants and acted as effective fertilisers accelerating the growth of protective vegetation. In this time, the vegetation gradually took over the protective function of the geotextiles, and once they were completely destroyed, the plants ensured the effective protection of the slopes.

Author details

Jan Broda
University of Bielsko-Biala, Faculty of Materials, Civil and Environmental Engineering, Institute of Textile Engineering and Polymer Materials, Bielsko-Biala, Poland

*Address all correspondence to: jbroda@ath.bielsko.pl

References

[1] Johnson NAG, Wood EJ, Ingham PE, et al. Wool as a technical fibre. Journal of the Textile Institute. 2003;**94**:26-41

[2] Johnson NAG, Russell IM. Advances in Wool Technology. Cambridge: Woodhead; 2009

[3] Ballagh KO. Acoustical properties of wool. Applied Acoustics. 1996;**48**(2):101-120. DOI: 10.1016/0003-682X(95)00042-8

[4] Zach J, Korjenic A, Petránek V, Hroudova J, Bednar T. Performance evaluation and research of alternative thermal insulations based on sheep wool. Energy and Buildings. 2012;**49**:246-253. DOI: 10.1016/j.enbuild.2012.02.014

[5] Bosia D, Savia L, Thiebat F, Patrucco A, Fantucci S, Piccablotto G, et al. Sheep wool for sustainable architecture. Energy Procedia. 2015;**78**:315-320. DOI: 10.1016/j.egypro.2015.11.650

[6] Korjenic A, Klarić S, Hadžić A, Korjenic S. Sheep wool as a construction material for energy efficiency improvement. Energies. 2015;**8**: 5765-5781. DOI: 10.3390/en8065765

[7] Rey R, Uris A, Alba J, Candelas P. Characterization of sheep wool as a sustainable material for acoustic applications. Materials. 2017;**10**(11):1277. DOI: 10.3390/ma10111277

[8] Broda J, Baczek M. Acoustic properties of multi-layer wool nonwoven structures. Journal of Natural Fibers. Published On line: DOI 10.1080/15440478.2019.1584078

[9] Radetic MM, Jocic DM, Jovancic PM, Petrovic ZL, Thomas HF. Recycled wool-based nonwoven material as an oil sorbent. Environmental Science and Technology. 2003;**37**:1008-1012

[10] Radetic M, Ilic V, Radojevic D, Miladinovic R, Jocic D, Jovancic P. Efficiency of recycled wool-based nonwoven material for the removal of oils from water. Chemosphere. 2008;**70**(3):525-530. DOI: 10.1016/j.chemosphere.2007.07.005

[11] Radetic M, Jocic D, Jovancic P, Rajakovic L, Thomas H, Petrovic ZL. Recycled-wool-based nonwoven material as a sorbent for lead cations. Journal of Applied Polymer Science. 2003;**90**:379-386

[12] Kar P, Misra M. Use of keratin fiber for separation of heavymetals from water. Journal of Chemical Technology and Biotechnology. 2004;**79**:1313-1319. DOI: 10.1002/jctb.1132

[13] Muthu S. Sustainable Fibres and Textiles. Woodhead Publishing; 2017

[14] Kadam VV, Meena LR, Singh S, Shakyawar DB, Naqvi SMK. Utilization of coarse wool in agriculture for soil moisture conservation. Indian Journal of Small Ruminants. 2014;**20**:83-86

[15] Skudelny K. Schafwollmatten fur Erosionsschutz und Begrunung. Nachhaltigkeit bei Verkehrwegeboschungen. Verkehr Umwelt. 2012:36-40

[16] Ament R, Cuelho E, Pokorny M, Jennings S. Evaluation of Effectiveness and Cost-Benefits of Woolen Roadside Reclamation Products. Final Report. Montana State University – Bozeman; 2014. DOI: 10.13039/100009209

[17] Prambauer M, Wendeler C, Weitzenböck J, Burgstaller C. Biodegradable geotextiles—An overview of existing and potential materials. Geotextiles and Geomembranes. 2019;**47**:48-59

[18] McNeil SJ, Sunderland MR, Zaitseva LI. Closed-loop wool carpet recycling. Resources, Conservation and Recycling. 2017;**51**:220-224. DOI: 10.1016/j.resconrec.2006.09.006

[19] Zheljazkov VD, Stratton GW, Pincock J, Butler S, Jeliazkova EA, Nedkov NK, et al. Wool-waste as organic nutrient source for container-grown plants. Waste Management. 2009;**29**:2160-2164. DOI: DOI.org/10.1016/j.wasman.2009.03.009

[20] Böhme M, Pinker I, Grunerberg H. Sheep wool as fertilizer for vegetables and flowers in organic farming. Acta Horticulturae. 2012;**933**:195-202

[21] Gupta S, Anshumala S, Sarika S, Narindra B. Growth, macro and micronutrient concentration in cluster bean (cyamopsis tetragonoloba), plant tissue as well as in soil when amended with wool as fertilizer. Journal of Environmental Research and Development. 2014;**8**(3A):607-613

[22] Zoccola M, Montarsolo A, Mossotti R, Patrucco A, Tonin C. Green hydrolysis as an emerging technology to turn wool waste into organic nitrogen fertilizer. Waste and Biomass Valorization. 2015;**6**:891-897. DOI: 10.1007/s12649-015-9393-0

[23] Aymerich F, Fenu L, Meloni P. Effect of reinforcing wool fibres on fracture and energy absorption properties of an earthen material. Construction and Building Materials. 2012;**27**:66-72. DOI: 10.1016/j.conbuildmat.2011.08.008

[24] Galán-Marín C, Rivera-Gómez C, Petric-Gray J. Clay-based composite stabilized with natural polymer and fibre. Construction and Building Materials. 2010;**24**:1462-1468. DOI: 10.1016/j.conbuildmat.2010.01.008

[25] Galán-Marín C, Rivera-Gómez C, Petric-Gray J. Effect of animal fibres reinforcement on stabilized earth mechanical properties. Journal of Biobased Materials and Bioenergy. 2010;**4**(2):121-128. DOI: 10.1166/jbmb.2010.1076

[26] Blyskal B. Fungi utilizing keratinous substrates. International Biodeterioration and Biodegradation. 2009;**63**:631-653

[27] Korniłłowicz-Kowalska T, Bohacz J. Biodegradation of keratin waste: Theory and practical aspects. Waste Management. 2011;**31**:1689-1701

[28] Queiroga AC, Pintado ME, Malcata FX. Potential use of wool-associated bacillus species for biodegradation of keratinous materials. International Biodeterioration & Biodegradation. 2012;**70**:60-65

[29] Rippon J. The structure of wool. In: Lewis DM, Rippon JA, editors. The Coloration of Wool and other Keratin Fibres. 1st ed. SDC: Wiley; 2013

[30] Kunert J. Keratin decomposition by dermatophytes. II: Presence of S-sulfocystine and cysteic acid in soluble decomposition products. Zeitschrift für Allgemeine Mikrobiologie. 1976;**16**:97-105

[31] Ruffin O, Andrieu S, Biserte G, Biguet J. Sulphitolysis in keratinolysis. Biochemical proof. Sabouraudia: Journal of Medical and Veterinary Mycology. 1976;**14**:181-184

[32] Christeller JT. Degradation of wool by Hofmannophila pseudospretella (Lepidoptera: Oecophoridae) larval midgut extracts under conditions simulating the midgut environment. Archives of Insect Biochemistry and Physiology. 1995;**32**(2):99-119. DOI: 10.1002/(SICI)1520-6327(1996)33:2

[33] Yamamura S, Morita Y, Hasan Q, Yokoyama K, Tamiya E. Keratin degradation: A cooperative action of two

enzymes from Stenotrophomonas sp. Biochemical and Biophysical Research Communications. 2002;**294**:1138-1143

[34] Bradbury JH. In: Anfinsen CB Jr, Edsall JT, Richards FM, editors. Advances in Protein Chemistry. Vol. 27. New York, USA: Academic Press; 1973. p. 111

[35] Bradbury JH, Chapman GV, King NRL. Australian Journal of Biological Sciences. 1965;**18**:353

[36] Evans DJ, Lanczki M. Cleavage of integral surface lipids of wool by aminolysis. Textile Research Journal. 1997;**67**:435-444

[37] Hutchinson S, Evans D, Corino G, Kattenbelt J. An evaluation of the action of thioesterases on the surface of wool. Enzyme and Microbial Technology. 2007;**40**:1794-1800

[38] Harland DP, Caldwell JP, Woods JL, Walls RJ, Bryson WG. Arrangement of trichokeratin intermediate filaments and matrix in the cortex of merino wool. Journal of Structural Biology. 2011;**173**:29-37

[39] Robbins C. The cell membrane complex: Three related but different cellular cohesion components of mammalian hair fibers. Journal of Cosmetics Science. 2009;**60**:437-465

[40] Kunert J. Biochemical mechanism of keratin degradation by the actinomycete Streptomyces fradiae and the fungus Microsporum gypseum: A comparison. Journal of Basic Microbiology. 1989;**29**:597-604

[41] Kunert J, Krajci D. An electron microscopy study of keratin degradation by the fungus Microsporum gypseum in vitro. Mykosen. 1981;**24**(8):485-496

[42] Wilson AS, Dodson HI, Janaway RC, Pollard AM, Tobin DJ. Selective biodegradation in hair shafts derived from archaeological, forensic and experimental contexts. British Journal of Dermatology. 2007;**157**:450-457

[43] Filipello Marchisio V, Fusconi A, Rigo S. Keratinolysis and its morphological expression in hair digestion by airborne fungi. Mycopathologia. 1994;**127**:103-115

[44] Filipello Marchisio V. Keratinophilic fungi: Their role in nature and degradation of keratinic substrates. Revista Iberoamericana de Micología. 2000;**17**:86-92

[45] DeGaetano D, Kempton JB, Rowe WF. Fungal tunneling of hair from a buried body. Journal of Forensic Sciences. 1992;**37**(4):1048-1054

[46] Chizhov AS, Ol'khov AA, Monakhov TV, Shibryaev LS, Iordanskii AL. Thermooxidation and biodegradation of nonwoven biopolymer fibrous materials. Polymer Science, Series D. 2018;**11**(3):339-343. DOI: 10.1134/S1995421218030036

[47] Gochel M, Belly M, Knott J. Biodeterioration of wool during storage. International Biodeterioration and Biodegradation. 1992;**30**:77-85

[48] Lin Z, Cao L, Liu M, Xie J, Feng D, Peng Z, et al. Physical and chemical changes of wool fabrics after artificial ageing. Key Engineering Materials. 2015;**671**:59-64

[49] Solazzo C, Dyer JM, Clerens S, Plowman J, Peacock EE, Collins MJ. Proteomic evaluation of the biodegradation of wool fabrics in experimental burials. International Biodeterioration & Biodegradation. 2013;**80**:48-59. DOI: 10.1016/j.ibiod.2012.11.013

[50] Peacock EE. Biodegradation and characterization of water-degraded archaeological textiles created for conservation research. International Biodeterioration and Biodegradation. 1996;**38**:49-59. DOI: 10.1016/S0964-8305(96)00023-6

[51] Brandt L, Tranekjer L, Mannering U, Ringgaard M, Frei K, Willerslev E,

et al. Characterising the potential of sheep wool for ancient DNA analyses. Archaeological and Anthropological Sciences. 2011;**3**:209-221. DOI: 10.1007/s12520-011-0055-2

[52] Helbig R, Arnold R, Erth H, Roess T, Hevert W, Lischkowitz H. New Technologies for Manufacturing Extra Coarse Rope-like Biodegradable Geotextiles. Technical Textiles. 2006;**49**:185-187, 244-247

[53] PN-EN ISO 10319:2010. Geosyntetyki—Badanie wytrzymalosci na rozciaganie metoda szerokich probek

[54] PN-EN ISO 12236:2006. Geosyntetyki—Badanie na przebicie statyczne (badanie CBR)

[55] PN-EN ISO 13433:2006. Geotekstylia i wyroby pokrewne—Wyznaczanie dynamiki perforacji (badanie opadajacym stozkiem

[56] PN-ISO 7150-1:2002. Jakosc wody—Oznaczanie azotu amonowego—Czesc 1: Manualna metoda spektrometryczna

[57] PN-73C-04576/06. Woda i scieki. Badania zawartosci zwiazkow azotu. Oznaczanie azotu azotynowego metoda kolorymetryczna z kwasem sulfanilowym i d-naftyloamina.

[58] PN-82/C-04576/08:1982. Woda i scieki—Badania zawartosci zwiazkow azotu. Oznaczanie azotu azotanowego metoda kolorymetryczna z salicylanem sodowym

[59] PN-EN 25663:2001. Jakosc wody. Oznaczanie azotu Kjeldahla. Metoda po mineralizacji z selenem.

[60] Broda J, Przybylo S, Kobiela-Mendrek K, Binias D, Rom M, Grzybowska-Pietras J, et al. Biodegradation of sheep wool geotextiles. International Biodeterioration & Biodegradation. 2016;**115**:31-38. DOI: DOI.org/10.1016/j.ibiod.2016.07.012

[61] Broda J, Gawlowski A, Laszczak R, Mitka A, Przybylo S, Rom M, et al. Application of innovative meandrically arranged geotextiles for the protection of drainage ditches in the clay ground. Geotextiles and Geomembranes. 2017;**45**:45-53. DOI: DOI.org/10.1016/j.geotexmem.2016.07.003

[62] Broda J, Gawlowski A, Przybylo S, Binias D, Rom M, Grzybowska-Pietras J, et al. Innovative wool geotextiles designed for erosion protection. Journal of Industrial Textiles. DOI: 10.1177/1528083717695837

[63] Broda J, Gawlowski A, Grzybowska-Pietras J, Rom M, Przybylo S, Laszczak R. Application of geotextiles for the stabilization of steep slopes in gravel pits. Ecological Engineering. 2017;**18**: 71-77. DOI: 10.12912/23920629/66987

[64] Broda J, Przybylo S, Gawlowski A, Grzybowska-Pietras J, Sarna E, Rom M, et al. Utilisation of textile wastes for the production of geotextiles designed for erosion protection. The Journal of the Textile Institute. On line. DOI: 10.1080/00405000.2018.1486684

[65] Broda J, Gawlowski A. Influence of sheep wool on slope greening. The Journal of Natural Fibers. On line. DOI: 10.1080/15440478.2018.1534190

Chapter 7

Adsorptive Removal of Fluoride onto Different Waste Materials: Orange Juice Residue, Waste Seaweed, and Spent Cation-Exchange Resin

Katsutoshi Inoue, Hari Paudyal, Hidetaka Kawakita and Keisuke Ohto

Abstract

To effectively use waste materials in developing a sustainable society, adsorbents for removing trace or low concentrations of fluoride, which is difficult to be removed by conventional techniques, were prepared from three waste materials: orange juice residue, waste sea weed, and spent cation exchange resin. These adsorbents were loaded with tri- or tetravalent metal ions such as iron(III) and zirconium(IV), of which zirconium(IV) was found to be most suitable as the loaded metal ion. From the pH effect on adsorption, the adsorption mechanism was inferred, and adsorption and desorption was found to be controlled by changing pH values. The maximum adsorption capacities on zirconium(IV)-loaded orange juice residue, waste sea weed, and spent cation exchange resin were evaluated as 33.1, 18.1, and 37.6 mg/g, respectively, which were higher than those of most other adsorbents reported in literatures. They exhibited high selectivity for fluoride over other anionic species and high durability. Tests to remove trace concentrations of fluoride from actual waste plating solutions revealed that the concentration could be reduced below the acceptable level using small amounts of these adsorbents, i.e., it was reduced lower than 1.5 mg/dm3 (WHO standard) by adding 1 g of the adsorbents into 1 dm3 test solution.

Keywords: adsorptive removal of fluoride, waste materials, orange juice residue, waste seaweed, spent cation-exchange resin, zirconium(IV)-loaded adsorbents

1. Introduction

Fluoride is contained in effluents from various industries, such as plating, productions of silicon semiconductors and solar panels, nuclear fuels, and phosphoric acid from phosphate rocks, as well as the refining of niobium and tantalum. The conventional technique for removing fluoride from such effluents is by precipitation as calcium fluoride by adding calcium, such as in the form of lime; however, because of the low solubility product of calcium fluoride

IntechOpen

($Ksp = 3.9 \times 10^{-11}$ (e.g., see [1])), it is theoretically difficult to lower the concentration to less than 8 mg dm^{-3} and practically difficult to achieve less than 10 mg dm^{-3}. The maximum acceptable limit of fluoride in drinking water is 1.5 mg dm^{-3} (e.g., see [2]). Consequently, some additional treatment is necessary to meet the regulation for fluoride before discharging industrial waste solutions into the environment. Adsorption techniques, including ion-exchange, are most suitable to remove such trace concentrations of contaminants in water. According to the well-known Hoffmeister's selectivity series of anion-exchange [3], the selectivity of usual anion-exchange resins for fluoride over other anionic species, such as sulfate and chloride (which usually coexist together with fluoride), is low, so it is difficult to selectively remove trace concentration of fluoride from actual solutions using usual anion-exchange resins. It has therefore been proposed to use cation-exchange resins loaded with high-valency metal ions, such as tri- and tetravalent metal ions [4]. **Figure 1** shows, as an example, the mechanism of adsorption of arsenate on a sulfonic acid-type cation-exchange resin loaded with trivalent ferric ion. In this case, all three positive charges of the ferric ion cannot be neutralized only by the sulfonic acid functional groups ($-SO_3^-$) due to strong steric hindrance of the cross-linked polymer matrix; one or two positive charges are neutralized by anionic species present in aqueous solution, such as hydroxyl ions, which are substituted by the anionic species in question, such as arsenate or fluoride anions. Consequently, the adsorption and desorption of these anionic species are controlled by the pH values of the aqueous solution.

However, ion-exchange resins, which are highly porous materials, suffer from some drawbacks. One disadvantage is clogging of micropores of the resins by fine particles present in actual solutions, making the smooth operation in packed columns difficult; another issue is cracking of the resin, caused by numerous cycles of fluctuations in pH and solute concentrations over long periods of operation, like weathering of rocks. In such cases, the resins are wasted.

In our research, we attempted to develop an inexpensive removal technique for fluoride that is free from the above-mentioned troubles by using waste materials. In this work, biomass wastes such as orange juice residue [5, 6] and waste seaweeds [7] and spent cation-exchange resins [8] were employed. This study also aimed to find effective uses of waste materials to support the sustainable development of near-future society.

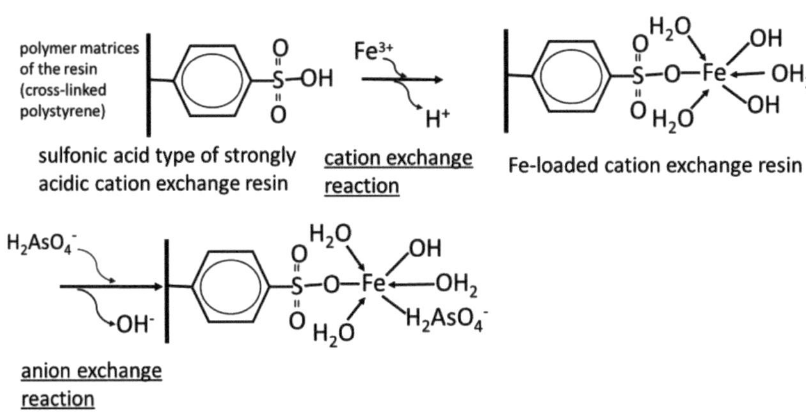

Figure 1.
Mechanism of adsorption of anionic species on cation-exchange resins loaded with high-valency metal ions (example of adsorption of arsenate ion on sulfonic acid type of strongly acidic cation-exchange resin loaded with ferric ion).

2. Preparation of adsorbents for fluoride removal

2.1 Preparation of adsorbents from orange juice residue

Orange juice is produced from whole oranges by mechanical peeling followed by juicing processes. Approximately, half of each orange is left as orange juice residue, abbreviated as OJR hereafter, which consists of thick outer rind, thin inner peel, seeds, and calyces. Currently, OJR is used as cattle feed and fertilizer in Japan. Its main components are polysaccharides, such as cellulose and pectin, partly methyl-esterified pectic acid, the chemical structures of which are shown in **Figure 2**, together with alginic acid, which is mentioned in Section 2.2.

To prepare the adsorption gel and improve its adsorption capacity for metal ions, OJR was treated with calcium hydroxide to increase the content of carboxylic acid groups, which are the adsorption sites for cationic metal ions. Saponification of the methyl-ester parts of pectin was carried out according to the reaction shown in **Figure 3** [5].

For example, about 100 g OJR obtained just after juicing, kindly donated by JA Saga Beverage Co. Ltd., was added to a juicer with 8 g calcium hydroxide, and the sample was pulverized. The resulting mixture was transferred to a beaker, along with a large volume of water, and the mixture was stirred for 24 h at 200 rpm and room temperature to facilitate the saponification reaction. The initial pH of the mixture was maintained at around 12.5 by adding sodium hydroxide solution. After decantation, the mixture was repeatedly washed using water and decanted until

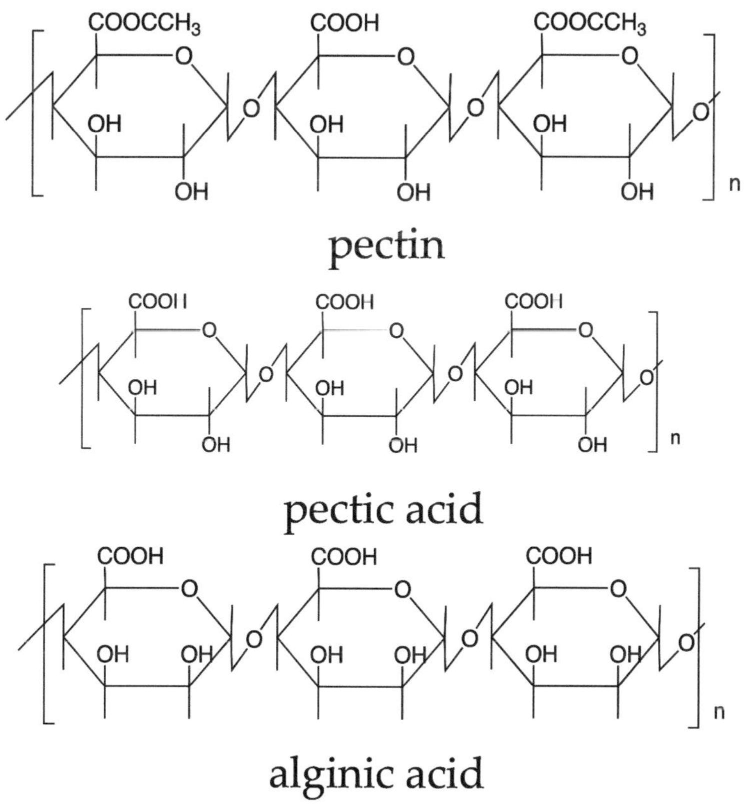

Figure 2.
Chemical structures of pectin, pectic acid, and alginic acid.

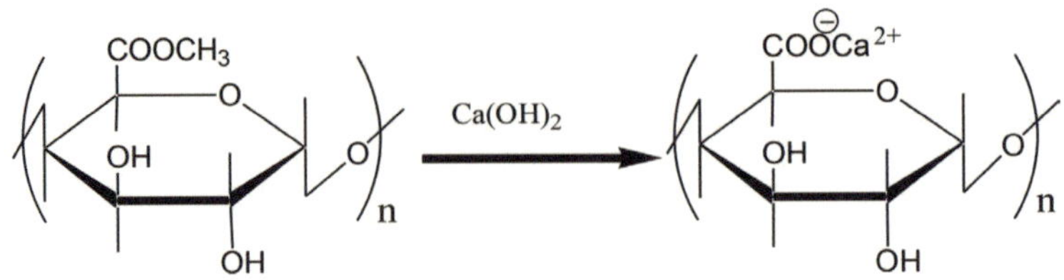

Figure 3.
Saponification of methyl-ester parts of pectin with calcium hydroxide.

it reached neutral pH. In this process, water washing is important because OJR contains citric acid, which functions as a water-soluble chelating agent, impeding adsorption of metal ions or eluting the loaded metal ions. Finally, the mixture was filtered to obtain a wet gel, which was dried in a convection oven for about 48 h at 70°C to obtain dry gel. The prepared dry gel is referred to as saponified OJR gel (SOJR gel). The specific surface area of this gel was measured as 7.25 $m^2 g^{-1}$. **Figure 4** shows the outline of the preparation route of SOJR gel and metal-loaded SOJR gel (zirconium(IV)-loaded SOJR gel, in this case), which will be mentioned later.

Figure 5 shows an image of scanning electron microscope of thus prepared SOJR gel. As seen from this figure, SOJR gel is not a porous material.

2.2 Preparation of adsorbents from waste seaweeds

In recent years, big cities adjacent to seas in Japan have been suffering from pollution of their sea sides by waste seaweeds, *Ulva japonica* in particular, caused by their overgrowth due to the eutrophication of the seas around these cities. Large amount of these seaweeds decay on beaches in summer and generating stench; consequently, the treatment of the seaweed wastes has become a serious environmental problem for municipalities of these coastal areas. The development of effective uses of these waste seaweeds is strongly required in Japan. The *Ulva* species of seaweed contains a high content of polysaccharides, including alginic acid (15–65%), lignin

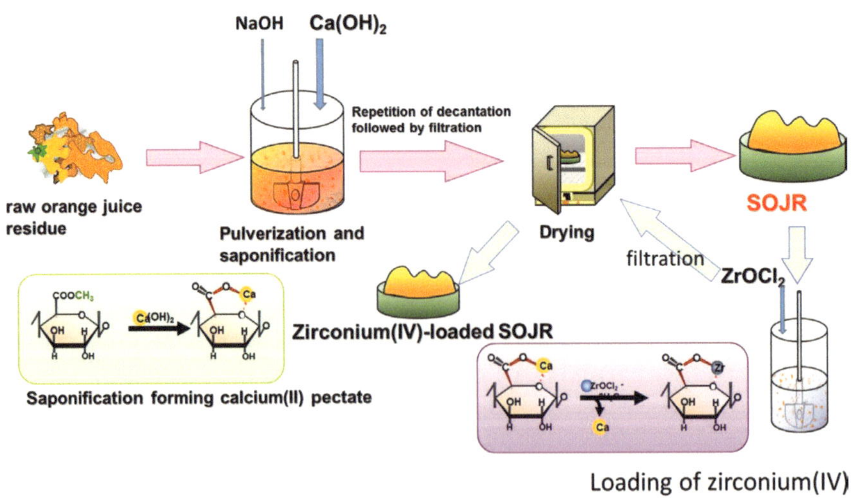

Figure 4.
Preparation route of adsorption gels from OJR.

Figure 5.
Scanning electron microscope image of SOJR gel.

(1–3%), protein (6–15%), lipid (1–4%), other amino acids, minerals, and water-soluble organics [9]. Alginic acid presents as calcium alginate with an egg-box structure.

In the present work [7], a sample of *Ulva japonica* collected at Wajiro coast in Fukuoka city, Japan, was washed several times with tap water followed by distilled water to remove impurities. It was pulverized after drying in a convection oven at 70°C for 24 h and sieved to a particle size of <100 μm. The material was then cross-linked with calcium chloride to lower its aqueous solubility. The cross-linked product was washed several times with distilled water and dried in a convection oven at 70°C. Thus, prepared waste seaweed gel is termed as cross-linked seaweed and abbreviated as CSW, hereafter. Similar to the case of SOJR gel, mentioned in Section 2.1, high-valency metal ions, such as zirconium(IV), were loaded on CSW to develop adsorption sites for fluoride, as shown in **Figure 6**.

2.3 Preparation of adsorbents from spent cation-exchange resin

Ion-exchange resins are extensively used in a variety of field, such as water softening, wastewater treatment, precious metal recovery, food industries, and

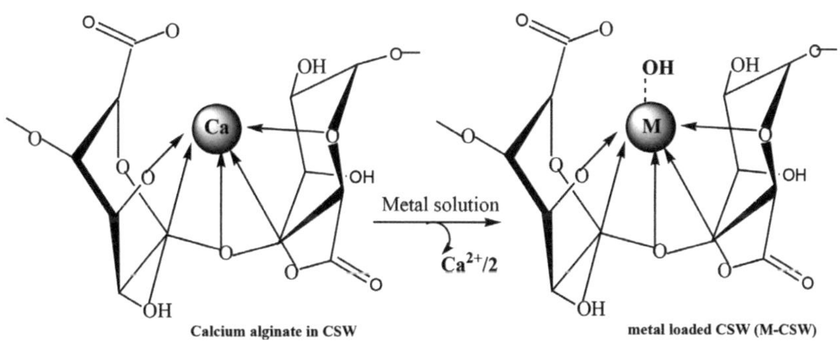

Figure 6.
Preparation of metal-loaded cross-linked seaweed (CSW) gel for fluoride removal [M = Zr(IV), La(III)].

catalytic processes. Although ion-exchange resins can be regenerated numerous times, they gradually deteriorate owing to the generation of many cracks, which impede smooth column operation and become unusable after many cycles of adsorption followed by elution as mentioned earlier. Tons of spent ion-exchange resins are discarded every year all over the world. Although they are unsuitable for column operation, their functional groups are still active; however, ideas for effective reutilization of such spent ion-exchange resins have not been proposed to date.

In our study [8], we attempted to effectively use such spent ion-exchange resins by pulverizing them into fine powders for use in batch operation together with a coagulating agent, that is, by means of the combined processes of adsorption and coagulation. A sample of spent cation-exchange resin containing sulfonic acid functional groups was pulverized into fine powders and converted to an anion exchanger by loading Zr(IV) ions for fluoride ion removal. The feasibility of using spent resin powder after loading Zr(IV) as ligand exchange-type material for fluoride adsorption was tested in combination with a coagulating agent in a batch system.

A sample of spent strongly acidic cation-exchange resin, MUROMAC MBX8-WH, which had been employed for the production of ion-exchange water, was kindly supplied by Muromachi Chemical Co. Ltd., Omuta, Japan. The polymer matrix of this resin is polystyrene cross-linked by divinylbenzene and it contained functional groups of sulfonic acid. The sample was pulverized into a fine powder, the average diameter of which was less than 20 μm. Thus, prepared resin powder is abbreviated as SRP hereafter. Because of the presence of an exchangeable hydrogen ion in the sulfonic acid group, waste resin powder can effectively adsorb cationic species of metal ions, such as Cu(II), Cd(II), Fe(III), La(III), and Zr(IV), whereas it has low affinity toward anionic species like fluoride ion. Similar to the cases of SOJR and CSW, the adsorption sites for fluoride anion were developed by loading Zr(IV) ion onto the SRP. Here, Zr(IV) was loaded by replacing the hydrogen ion from sulfonic acid functional groups of SRP by the cation-exchange mechanism as shown in **Figure 7**. The material obtained in this way is termed Zr(IV)-loaded spent resin powder and abbreviated as Zr SRP hereafter.

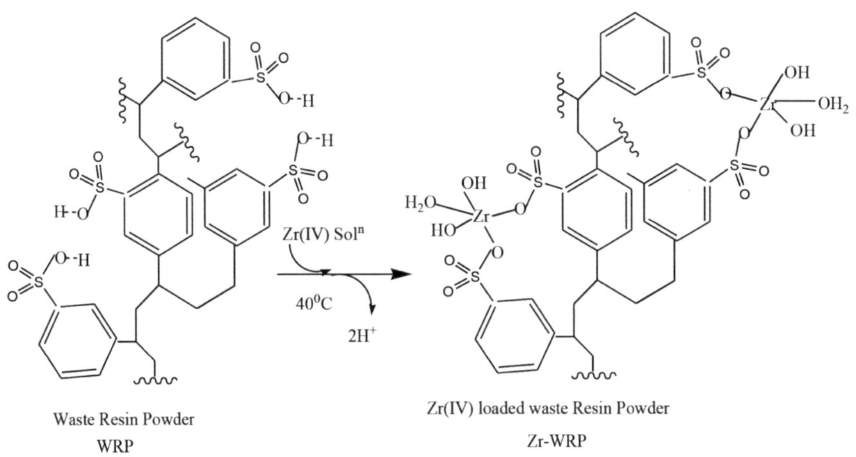

Figure 7.
Development of active sites for fluoride onto spent resin powder (SRP) containing sulfonic acid functional group via. Zr(IV) loading.

DOI: http://dx.doi.org/10.5772/intechopen.80448

3. Adsorption behavior of fluoride on saponified orange juice residue gels loaded with various high-valency metal ions

Figure 8 shows the effect of equilibrium pH on the adsorption of fluoride on SOJR gels loaded with various high-valency metal ions. It is apparent that pH greatly influenced the adsorption in the presence of the different metal ions, although the adsorption decreases with increasing pH, regardless of the loaded metal ion. Of the metal ions tested, Zr(IV) exhibits the most effective adsorption, at low pH in particular. The decrease in adsorption at high pH is attributed to the effect of hydroxyl ions, which compete with fluoride ion for adsorption sites on SOJR gel, as depicted in **Figure 9** for the case of Zr(IV)-loaded SOJR gel.

Also from this figure, it is apparent that the adsorption and desorption of fluoride can be easily controlled by changing pH.

Figure 10 shows the adsorption isotherms of fluoride, which is the relationship between the amount of adsorbed fluoride (q) and the fluoride concentration (Ce) in the aqueous solution at equilibrium, on SOJR gels loaded with various high-valency metal ions at 30°C. It is evident that SOJR gel loaded with Zr(IV) exhibited the highest adsorption capacity for fluoride. Consequently, we used Zr(IV)-loaded adsorbent in subsequent work.

Figure 11 shows the effect of coexisting species on the adsorption of fluoride. Although sulfate most effectively interferes with the adsorption of fluoride, it is only less than 2%; consequently, the effect of coexisting species was practically negligible, that is, Zr(IV)-loaded SOJR gel exhibited high selectivity to fluoride over coexisting species. Similar was observed also in the adsorption on Zr(IV)-loaded other adsorbents.

Figure 12 shows time variation in the adsorption of fluoride on Zr(IV)-loaded SOJR gel at various temperatures. Equilibrium was attained within a short contact time (shorter than 100 min), indicating rapid adsorption, regardless of temperature.

Figure 13 shows the effect of initial pH on the elution of fluoride adsorbed on Zr(IV)-loaded SOJR gel. The adsorbed fluoride was quantitatively eluted at pH higher than 12.

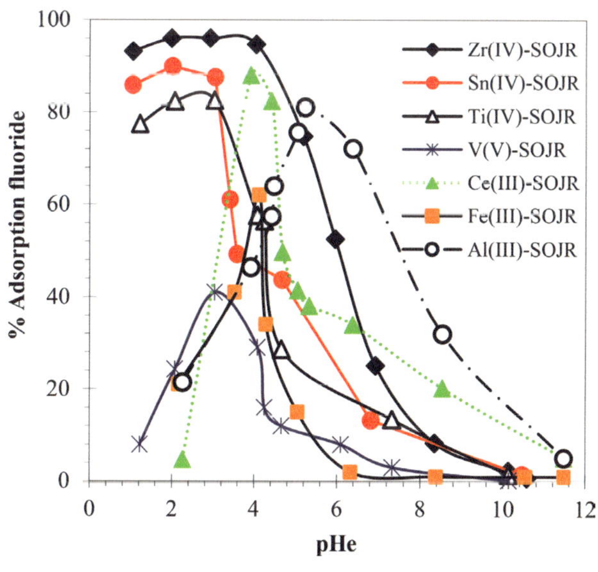

Figure 8.
Effect of equilibrium pH (pHe) on the adsorption of fluoride on saponified orange juice residue (SOJR) gels loaded with various high-valency metal ions.

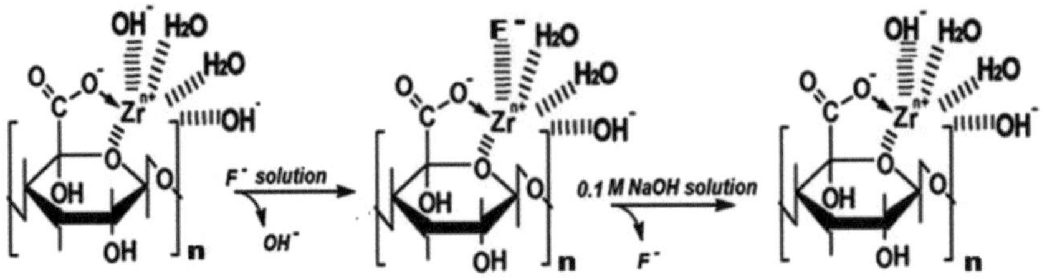

Figure 9.
Adsorption and desorption mechanism of fluoride on Zr(IV)-loaded saponified orange juice residue (SOJR) gels.

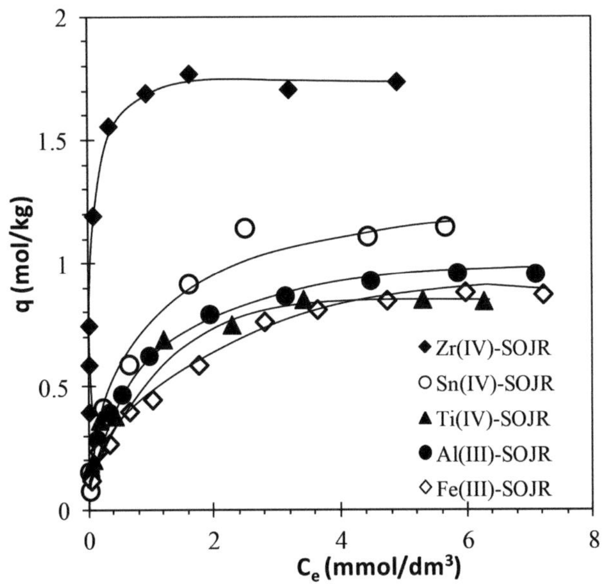

Figure 10.
Adsorption isotherms of fluoride on saponified orange juice residue (SOJR) gels loaded with various high-valency metal ions at 30°C.

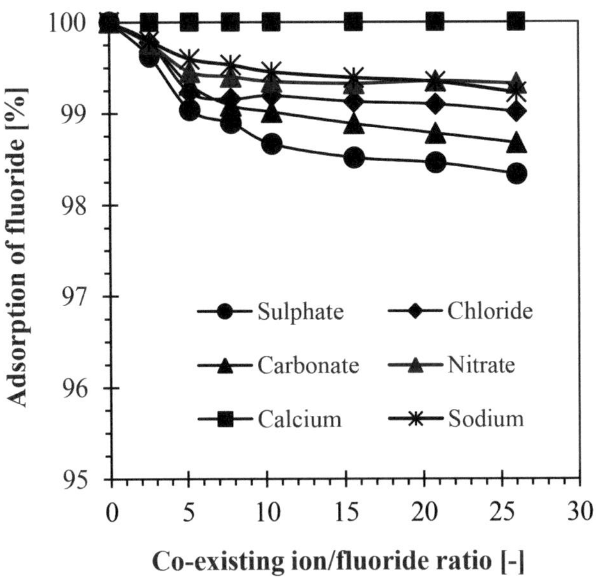

Figure 11.
Effect of coexisting species in the aqueous solution on adsorption of fluoride on Zr(IV)-loaded saponified orange juice residue (SOJR) gel.

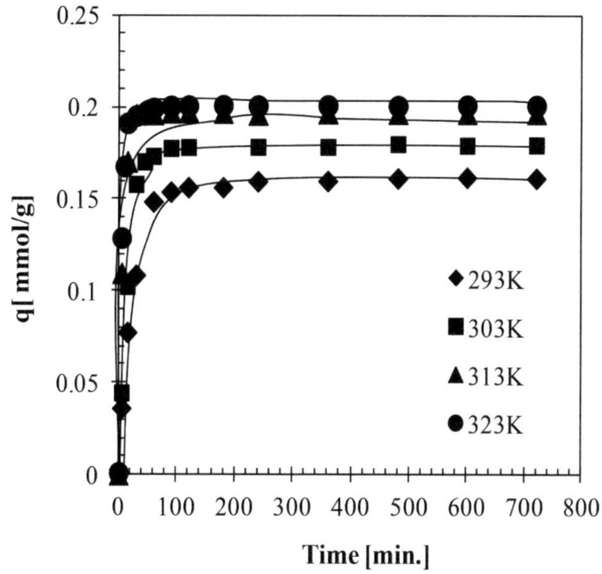

Figure 12.
Time variation in the amount of adsorption of fluoride (q) on Zr(IV)-loaded saponified orange juice residue (SOJR) gel at various temperatures.

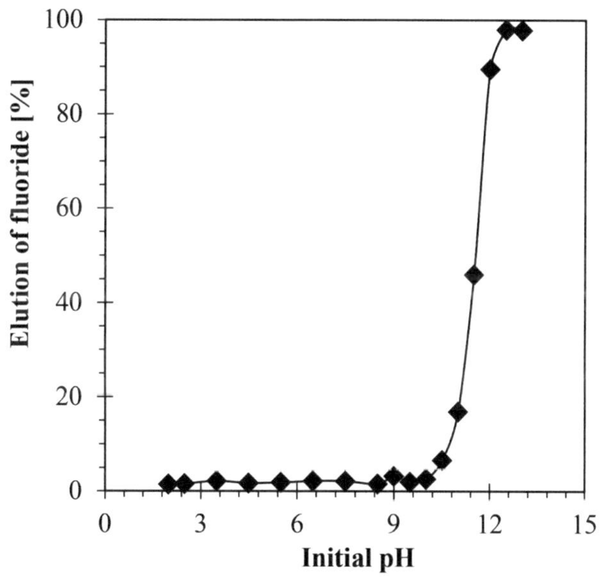

Figure 13.
Effect of pH on elution of fluoride adsorbed on Zr(IV)-loaded saponified orange juice residue (SOJR) gel.

Figure 14 shows effect of initial fluoride concentration at various pH values on leakage of the loaded zirconium(IV) from the SOJR gel. Such leakage, which would cause degradation of the adsorbent, was practically negligible at all pH values, except below pH 3, where the loaded metal ions are replaced by hydrogen ions by cation-exchange mechanism.

Figure 15 shows the result of cycle test of adsorption of fluoride on Zr(IV)-loaded SOJR gel followed by elution using 0.1 mol dm^{-3} NaOH. It can be concluded that the adsorption and desorption behaviors of fluoride on this gel were not significantly degraded, even after 10 cycles of adsorption and elution. The Zr(IV)-loaded SOJR gel exhibited high durability.

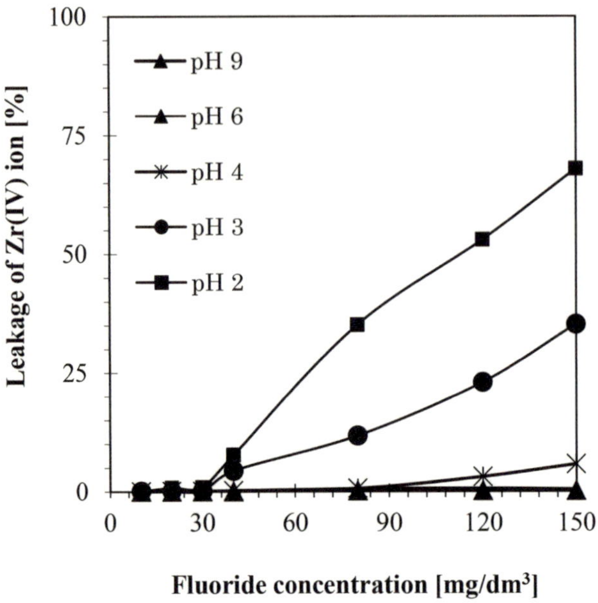

Figure 14.
Effect of initial fluoride concentration of fluoride at various pH values on leakage of loaded zirconium(IV) from saponified orange juice residue (SOJR) gel.

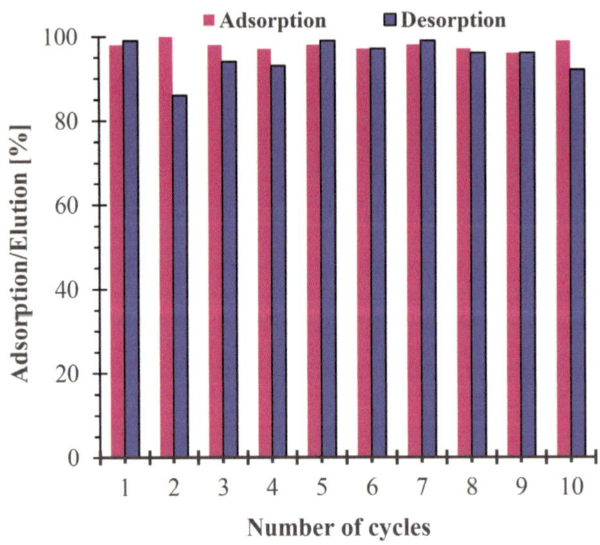

Figure 15.
Effect of number of cycles of adsorption of fluoride on Zr(IV)-loaded saponified orange juice residue (SOJR) gel followed by elution using 0.1 mol dm^{-3} NaOH.

4. Adsorption behavior of fluoride on waste seaweed gel loaded with various high-valent metal ions

Figure 16 shows the effect of equilibrium pH on the adsorption of fluoride on the cross-linked seaweed gel (CSW) and CSW gels loaded with Zr(IV) and La(III). Unloaded CSW gel also exhibits some adsorption of fluoride, which is attributable to calcium(II) ion contained in seaweed as calcium alginate. Similar to SOJR gel, the adsorption of fluoride was greatly improved by loading high-valency metal ions, of which Zr(IV) exhibited better adsorption than La(III). The decrease in adsorption

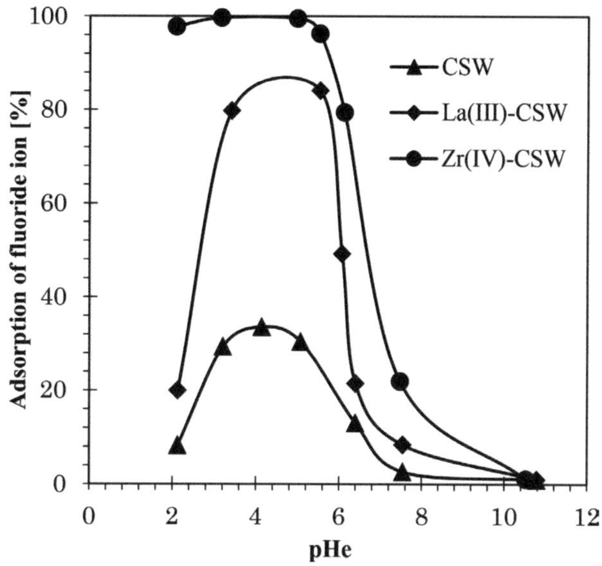

Figure 16.
Effect of equilibrium pH on the adsorption of fluoride on the cross-linked seaweed gel (CSW) and metal-loaded CSW gels [7].

with increasing pH at high pH is attributable to the competition of fluoride ion with hydroxyl ion for adsorption sites on the loaded metal ions; that is, the adsorption and desorption mechanism in this system can also be reasonably interpreted by that depicted in **Figure 9** for SOJR gel.

Figure 17 shows the effect of initial pH on leakage of the loaded metal ions. It is apparent that, compared with La(III), the loaded Zr(IV) is much more difficult to leak from CSW than La(III); that is, Zr(IV) is also more suitable as the loaded metal ion on CSW.

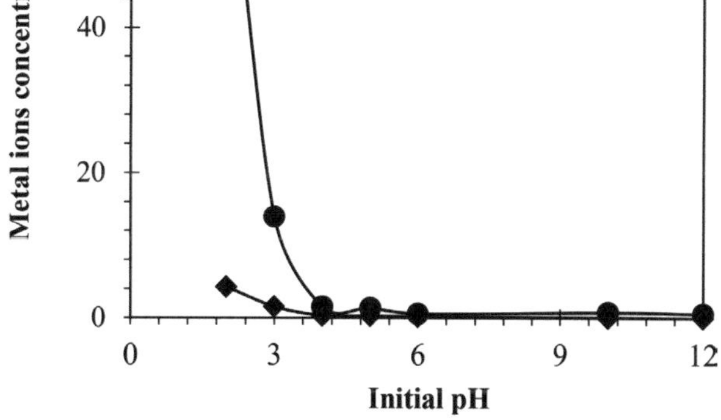

Figure 17.
Effect of initial pH on leakage of the loaded metal ions from cross-linked seaweed (CSW) gels.

5. Adsorption behavior of fluoride on pulverized spent cation-exchange resin loaded with zirconium(IV) ions

A small amount of commercially available inorganic coagulating agent, MAGNETITE III, the main components of which are calcium, aluminum, and iron, was added with Zr-SRP to test solutions containing trace concentration of fluoride to achieve smooth solid/liquid separation or smooth filtration. The coagulating agent is abbreviated as CA, hereafter. It was found that stable floc was formed at a CA/Zr-SRP ratio higher than 0.2.

Figure 18 shows the effect of equilibrium pH on the adsorption of fluoride onto SRP before and after Zr(IV) loading in the presence and absence of CA. Although SRP exhibited only negligible adsorption for fluoride, the adsorption was greatly

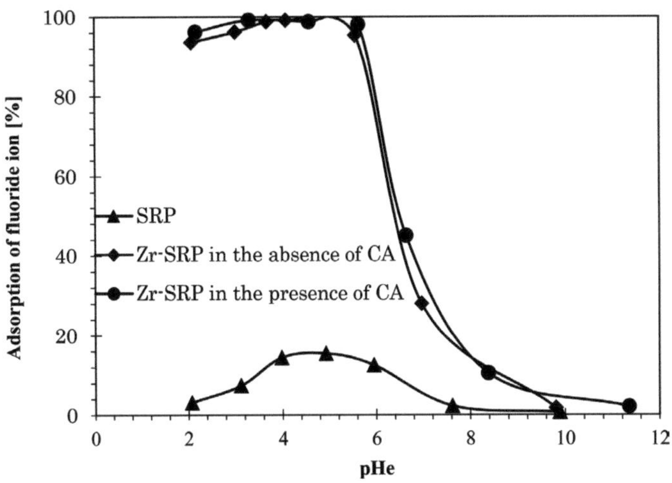

Figure 18.
Effect of equilibrium pH on adsorption of fluoride on pulverized spent cation-exchange resin (SRP) before and after Zr(IV) loading [8].

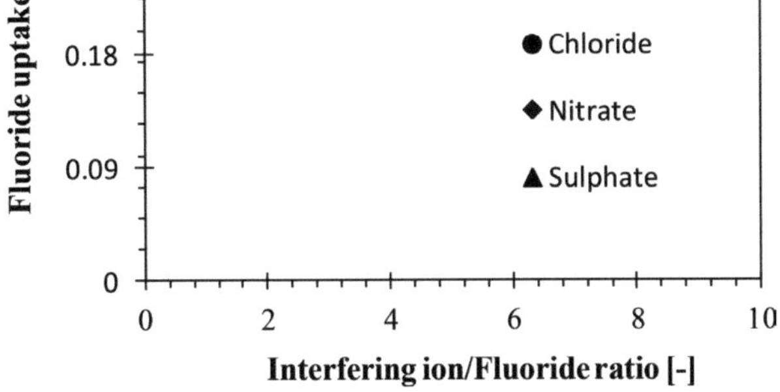

Figure 19.
Effect of interfering anionic species on adsorption of fluoride on Zr(IV)-loaded pulverized spent cation-exchange resin (Zr-SRP) [8].

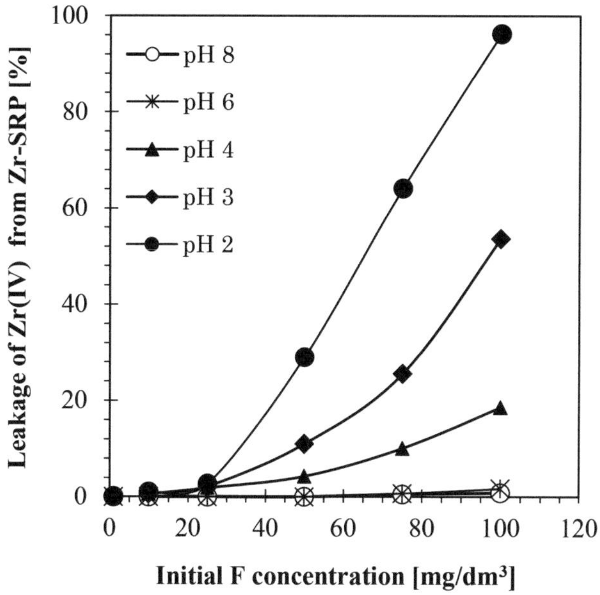

Figure 20.
Effect of fluoride concentration on leakage of loaded Zr(IV) from Zr(IV)-loaded pulverized spent cation-exchange resin (Zr-SRP) at various pH values [8].

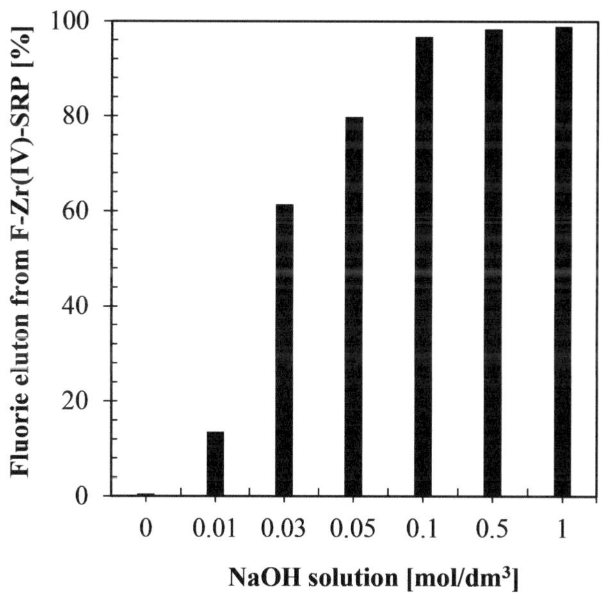

Figure 21.
Effect of concentration of sodium hydroxide on the elution of fluoride adsorbed on Zr(IV)-loaded pulverized spent cation-exchange resin (Zr-SRP) [8].

improved by loading Zr(IV), similar to the cases of SOJR and CSW. The small amount of adsorption on unloaded SRP at pH = 4–6 is considered to be due to calcium accumulated in the resin during its previous function in water treatment. Adsorption of fluoride also rapidly decreased with increasing pH at pH values above 6, similar to the cases of SOJR and CSW, which is attributable to the competition of fluoride ions with hydroxyl ions for the adsorption sites on Zr-SRP. Adsorption of fluoride was not affected by the addition of small amount of CA.

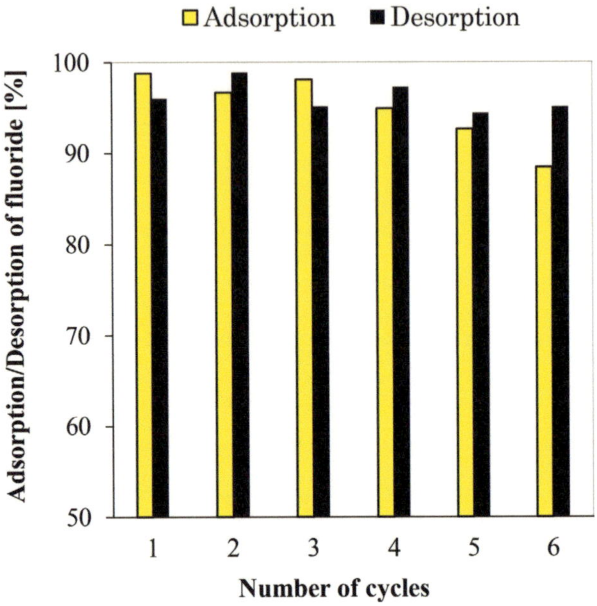

Figure 22.
Effect of number of cycles of adsorption of fluoride on Zr(IV)-loaded pulverized spent cation-exchange resin (Zr-SRP) followed by elution using 0.1 Mol dm^{-3} NaOH solution [8].

Figure 19 shows the effect of interfering anionic species (chloride, nitrate, and sulfate) present in aqueous fluoride solutions on the adsorption of fluoride on Zr-SRP. Also in this case, the effect of coexisting anionic species was relatively low.

Figure 20 shows leakage of Zr(IV) from Zr-SRP by varying concentration of fluoride at various pH values. It is evident that the leakage was negligible at pH values higher than 6, regardless of fluoride concentration.

Figure 21 shows effect of sodium hydroxide concentration on the elution of fluoride adsorbed on Zr-SRP. The results suggest that nearly quantitative elution of fluoride can be achieved by 0.1 mol dm^{-3} sodium hydroxide solution also in this case.

Figure 22 shows the result of durability test of Zr-SRP for repeating adsorption of fluoride followed by elution for up to 6 cycles. This figure suggests that Zr-SRP is not significantly degraded up to at least 6 cycles.

6. Comparison of adsorption capacities for fluoride on Zr(IV)-loaded adsorbents prepared from wastes

Adsorbents are evaluated in terms of their maximum adsorption capacities and selectivity for targeted species over other coexisting species. Selectivity for fluoride of three adsorbents prepared from wastes over other anionic species is discussed in the preceding sections. In this section, the maximum adsorption capacities on Zr(IV)-loaded SOJR, CSW, and SRP for fluoride were compared.

Figure 23 shows adsorption isotherms for fluoride on Zr(IV)-loaded SOJR, CSW, and SRP at 30°C. For all of these adsorbents, the adsorption amount increased with increasing fluoride concentration in the low concentration range and tends to approach a constant value for each adsorbent in the high concentration range, suggesting typical Langmuir-type adsorption.

These adsorption isotherms were therefore replotted according to the Langmuir equation, expressed by Eq. (1) as shown in **Figure 24**.

$$Ce/q = 1/q_{max}b + Ce/q_{max} \tag{1}$$

where q_{max} and b denote the maximum adsorption capacity and adsorption equilibrium constant, respectively, based on the Langmuir's monolayer adsorption theory.

As expected from Eq. (1), the plots in **Figure 24** lay on straight lines. From the values of intercepts of these straight lines with the ordinate and their slopes, the

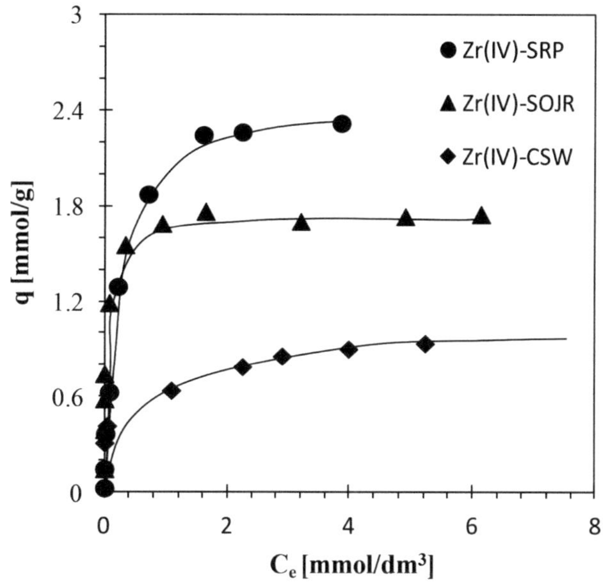

Figure 23.
Adsorption isotherms of fluoride on Zr(IV)-loaded saponified orange juice residue (SOJR), cross-linked seaweed (CSW) gels, and pulverized spent cation-exchange resin (SRP) at 30°C.

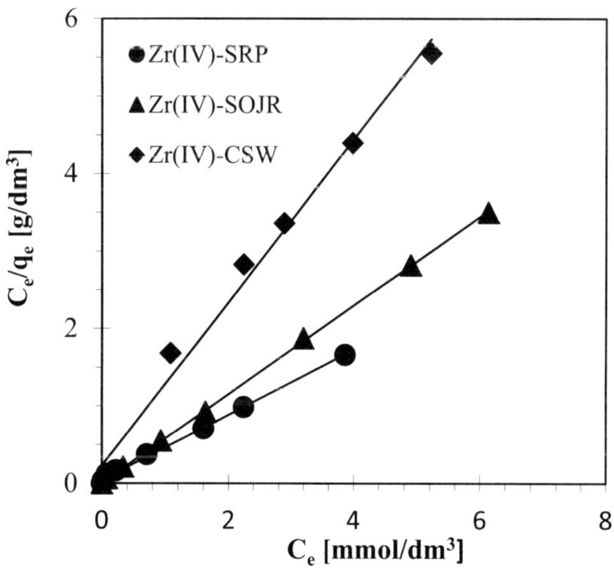

Figure 24.
Replots of **Figure 23** *according to Eq. (1), the Langmuir's equation.*

Adsorbents	q_{max} (mmol/g-dry weight)	b (dm^3/mmol)
Zr(IV)-loaded SOJR	1.74	3.11
Zr(IV)-loaded CSW	0.95	5.04
Zr(IV)-loaded SRP	1.98	1.18

Table 1.
Maximum adsorption capacities and adsorption equilibrium constants of fluoride on Zr(IV)-loaded saponified orange juice residue (SOJR) and cross-linked seaweeds (CSW) gels and pulverized spent cation-exchange resin (SRP) at 30°C.

Adsorbents	pH	q_{max} (mmol/g)	Temperature (°C)	Reference
Zr(IV)-loaded SOJR	4	1.74	30	[6]
Zr(IV)-loaded CSW	4	0.95	30	[7]
Zr(IV)-loaded SRP	4	1.98	30	[8]
Sn(IV)-loaded SOJR	3	1.18	30	[5]
Ti(IV)-loaded SOJR	3	0.93	30	[5]
Al(III)-loaded SOJR	6	1.03	30	[5]
Fe(III)-loaded SOJR	4	0.88	30	[5]
La(III)-loaded SOJR	4	1.06	30	[10]
Sc(III)-loaded SOJR	4	0.60	30	[10]
Sm(III)-loaded SOJR	5	1.22	30	[10]
Ho(III)-loaded SOJR	5	0.93	30	[10]
Ce(III)-loaded SOJR	4	0.91	30	[10]
Ce(IV)-loaded SOJR	4	1.53	30	[10]
READF(PG)	3	2.10	30	[11]
READF(HG)	3	2.35	30	[11]
Zr-impregnated cashew nut shell carbon	3	0.09	30	[12]
La-loaded 200CT resin	6	1.34	30	[13]
Indian FR 10	7	0.068	30	[14]
Ceralite IRA 400	7	0.078	30	[14]
La-loaded CL gelatin	5–7	1.12	25	[15]
Nd-modified chitosan	7	1.17	30	[16]
Fe-loaded cotton	4	0.97	25	[17]
Protonated chitosan	7	0.38	30	[18]
Activated alumina	5–6	0.86	30	[19]
AFB-Pr resin	5	0.026	30	[20]
La-impregnated alumina	5.7–8	0.35	Room temperature	[21]
Spirogyra species IO1	7	0.05	30	[22]
Waste mud	5	0.22	20	[23]
Nanoaluminum hydroxide	5.2	0.17	25	[24]
Magnesia/chitosan composite	10	0.11	Room temperature	[25]
Waste carbon slurry	7.5	0.23	25	[26]
Montmorillonite	6	0.01	25	[27]

Table 2.
Maximum adsorption capacities of fluoride on various adsorbents.

maximum adsorption capacities and adsorption equilibrium constants were evaluated, as listed in **Table 1**. The data indicate that Zr(IV)-loaded SRP exhibits the highest loading capacity.

These maximum adsorption capacities were compared with those on SOJR gel loaded with other high-valency metal ions and with those reported in literature, as listed in **Table 2**. Zr(IV)-loaded SRP and Zr(IV)-loaded SOJR exhibited higher adsorption capacities than other adsorbents, including Zr-containing adsorbents, except for READF, a commercially available porous resin that contains fine zirconium hydroxide powder in its pores.

7. Removal tests of trace concentrations of fluoride from actual waste plating solutions using Zr(IV)-loaded adsorbents prepared from wastes

On the basis of the results of fundamental adsorption tests using Zr(IV)-loaded SOJR, CSW, and SRP discussed above, adsorptive removal of trace concentrations of fluoride was tested from actual waste plating solutions using these adsorbents [6–8]. A sample of the actual waste plating solution was kindly donated by Federation of Electro Plating Industry Association, Japan, Tokyo, Japan. The donated sample solutions contained 12.2 mg/dm^3 of fluoride together with about 750 mg dm^{-3} of sulfate, as well as following cationic species (mg dm^{-3}): Al 13.8, Ca 10.5, Fe 36.4, Cu 9.8, and Zn 1.9. Its pH value was 6.7. **Figure 25** shows the plots of fluoride concentration after the treatment using these three adsorbents as a function of S/L ratio, defined as the ratio of dry mass of added adsorbent to volume of treated solution together with the permissible level for fluoride in effluents in Japan (8 mg/dm^3), WHO standard for drinking water (1.5 mg/dm^3), and that in Japan (0.8 mg/dm^3). This figure indicates that fluoride concentration can be successfully lowered by adding small amount of adsorbents. The concentration was reduced to less than 0.8 mg/dm^3, the Japanese standard for fluoride, by adding only 2 g of adsorbents to 1 dm^3 of tested sample solution.

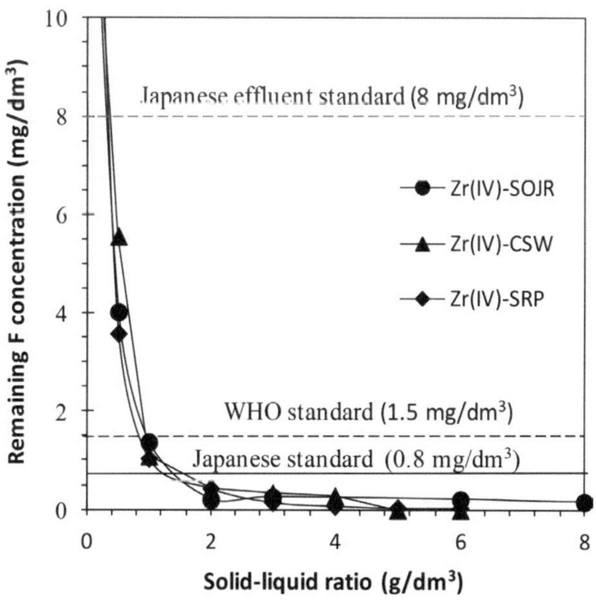

Figure 25.
Adsorptive removal of trace concentrations of fluoride from actual waste plating solution using three adsorbents prepared in the present work, showing relationship between concentration of fluoride remained in aqueous solution after adsorption and mass of adsorbent added to unit volume of solution.

8. Conclusion

For the purpose of removing trace or low concentrations of fluoride from aquatic environments, novel adsorbents were prepared from three waste materials, orange juice residue, waste seaweed, and pulverized spent cation-exchange resin containing sulfonic acid functional groups, by loading with tri- or tetravalent metal ions, such as iron(III), aluminum(III), rare earths(III), and zirconium(IV). Here, the loaded metal ions function as adsorption sites and waste materials function as polymer matrices immobilizing the metal ions. The adsorption mechanisms on these adsorbents were inferred from the results of adsorption tests from aqueous solutions at various pH values. Adsorption and desorption were found to be easily controlled by changing pH. Of the high-valency metal ions tested, zirconium(IV) was found to be the most suitable as the loaded metal ion because zirconium(IV)-loaded adsorbents exhibited highest adsorption of the tested metal-loaded adsorbents in the adsorption isotherm tests. Zirconium(IV)-loaded adsorbents exhibited high selectivity for fluoride over other anionic species present in aqueous solutions. Leakage of the loaded zirconium(IV) ions was found to be negligible, except at low pH, and high durability of these adsorbents was confirmed from repeated cycle tests of adsorption followed by desorption. The maximum adsorption capacities for fluoride on these adsorbents were higher than those of most adsorbents reported in literature. The removal test of trace concentration of fluoride from actual waste plating solutions revealed that the fluoride concentration could be lowered below the acceptable level by using only small amounts of these adsorbents; for example, it was reduced to less than 0.8 mg/dm^3, the Japanese standard for fluoride, by adding only 2 g of adsorbents to 1 dm^3 of tested sample solution. From these results, it can be expected that trace concentration of fluoride is effectively removed by using adsorbents produced from various waste materials at cheap cost.

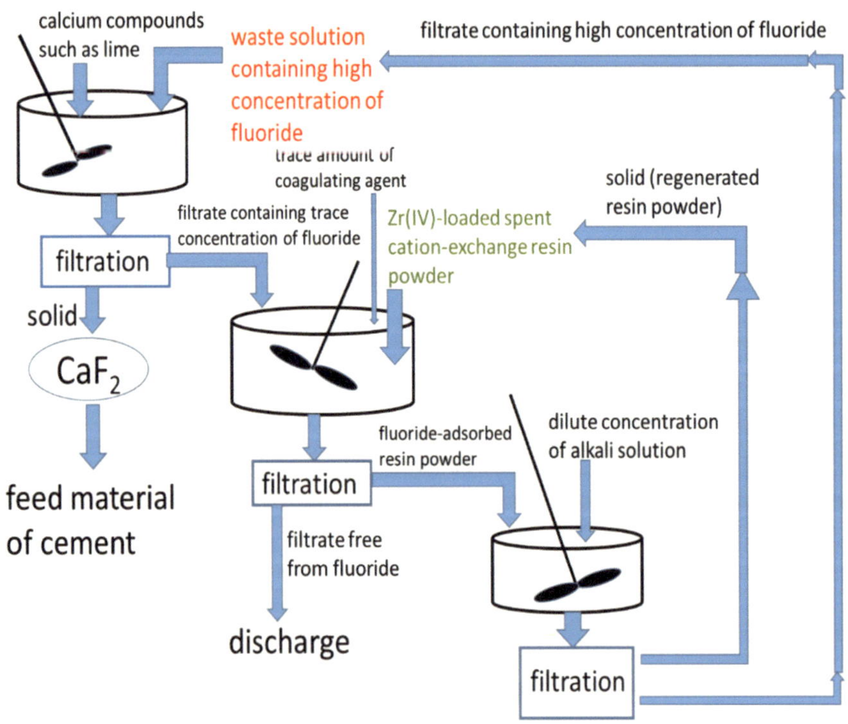

Figure 26.
Flow sheet for recovery of fluoride using powder of spent cation-exchange resin loaded with zirconium(IV).

Although it is not noticed in laboratory-scale tests and the small-scale adsorption tests mentioned in this study, the adsorbent prepared from orange juice residue was found to generate considerable stench at larger scale, such as pilot-scale operation. Taking account of this, the adsorbent prepared from spent cation-exchange resins is recommended to be employed in commercial operations. A process for the recovery of fluoride using this adsorbent is proposed as shown in **Figure 26**.

Acknowledgements

The present work was conducted in part in the project of Water-Saving Recycling Systems in 2012–2013 supported by New Energy and Industrial Development Technology Organization (NEDO). The authors acknowledge the kind donation of sample of waste plating solution by Federation of Electro Plating Industry Association, Japan, and those of pulverized spent cation-exchange resin powder and orange juice residue by Muromachi Chemical Co. Ltd., Omuta, Japan, and JA Saga Beverage, Saga, Japan, respectively.

Author details

Katsutoshi Inoue[1*], Hari Paudyal[1,2], Hidetaka Kawakita[1,2] and Keisuke Ohto[1]

1 Department of Applied Chemistry, Faculty of Science and Engineering, Saga University, Saga, Japan

2 Central Department of Chemistry, Tribhuvan University, Kathmandu, Nepal

*Address all correspondence to: kanoko1921@gmail.com

References

[1] Macaskill JB, Bates RG. Solubility product of calcium fluoride. The Journal of Physical Chemistry. 1977;**81**:496-498. DOI: 10.1021/j100520a026

[2] WHO. Guidelines for Drinking Water Quality, WHO, Geneva, Switzerland; 2004

[3] Helfferich F. Ion Exchange. McGraw-Hill; New York; 1962. pp. 250-323

[4] Yoshida I, Ueno K, Kobayashi H. Selective separation of arsenic (III) and (V) ions with ferric complex of chelating ion exchange resin. Separation Science and Technology. 1978;**13**:173-184. DOI: 10.1080/01496397808057099

[5] Paudyal H, Pangeni B, Inoue K, Kawakita H, Ohto K, Harada H, Alam S. Adsorptive removal of fluoride from aqueous solution by using orange waste loaded with multivalent metal ions. Journal of Hazardous Materials. 2011;**192**:676-682. DOI: 10.1016/j.jhazmat.2011.05.070

[6] Paudyal H, Pangeni B, Inoue K, Matsueda M, Suzuki R, Kawakita H, Ohto K, Biswas BK, Alam S. Adsorption behavior of fluoride ions on zirconium(IV)-loaded orange waste gel from aqueous solution. Separation Science and Technology. 2012;**47**:96-103. DOI: 10.1080/01496395.2011.607204

[7] Paudyal H, Pangeni B, Inoue K, Kawakita H, Ohto K, Ghimire KN, Alam S. Preparation of novel alginate based anion exchanger from *Ulva japonica* and its application for the removal of trace concentrations of fluoride from water. Bioresource Technology. 2013;**148**:221-227. DOI: 10.1016/j.biortech.2013.08.116

[8] Paudyal H, Inoue K, Kawakita H, Ohto K, Kamata H, Alam S. Removal of fluoride by effectively using spent cation exchange resin. Journal of Material Cycles and Waste Management. 2017;**20**:975-984. DOI: 10.1007/s10163-017-0659-4

[9] Kraan S. Algal Polysaccharides. Novel Application and Outlook. Galway: Ocean Harvest Technology Ltd.; 2012. pp. 489-532

[10] Paudyal H, Pangeni B, Ghimire KN, Inoue K, Ohto K, Kawakita H, Alam S. Adsorption behavior of orange waste gel for some rare earth ions and its application to the removal of fluoride from water. Chemical Engineering Journal. 2012;**195-196**:289-296. DOI: 10.1016/j.cej.1012.04.061

[11] Paudyal H, Pangeni B, Inoue K, Kawakita H, Ohto K, Alam S. Removal of fluoride from aqueous solution by using porous resins containing hydrated oxide of cerium(IV) and zirconium(IV). Journal of Chemical Engineering of Japan. 2012;**45**:331-336. DOI: 10.1252/jcej.11we181

[12] Algagumuthu G, Rajan M. Equilibrium kinetics of adsorption of fluoride on to zirconium-impregnated cashew nut shell carbon. Chemical Engineering Journal. 2010;**158**:451-457. DOI: 10.1016/j.cej.2010.01.017

[13] Fang L, Ghimire KN, Kuriyama M, Inoue K, Makino K. Removal of fluoride using some lanthanum loaded adsorbents with different functional groups and polymer matrices. Journal of Chemical Technology and Biotechnology. 2003;**78**:1038-1047. DOI: 10.1002/jctb.902

[14] Meenakshi S, Viswanathan N. Identification of selective ion exchange resin for fluoride adsorption. Journal of Colloid and Interface Science. 2007;**308**:438-450. DOI: 10.1016/j.jcis.2006.12.032

[15] Zhou Y, Yu C, Shan Y. Adsorption of fluoride from aqueous solution on

La^{3+}-impregnated cross-linked gelatin. Separation and Purification Technology. 2004;**36**:89-94. DOI: 10.1016/S1383-5866(03)00167-9

[16] Yao R, Meng F, Zhang L, Ma D, Wang M. Defluoridation of water using neodymium-modified chitosan. Journal of Hazardous Materials. 2009;**165**:454-460. DOI: 10.1016/j.jhazmat.2008.10.052

[17] Zhao Y, Li X, Liu L, Chen F. Fluoride removal by Fe(III)-loaded ligand exchange cotton cellulose adsorbent from drinking water. Carbohydrate Polymers. 2008;**72**:144-150. DOI: 10.1016/j.carbpol.2007.07.038

[18] Viswananthan N, Sundaram CS, Meenakshi S. Removal of fluoride from aqueous solution using protonated chitosan beads. Journal of Hazardous Materials. 2009;**161**:423-430. DOI: 10.1016/j.jhazmat.2008.03.115

[19] Ku Y, Chiou HM. The adsorption of fluoride ion from aqueous solution by activated alumina. Water, Air, and Soil Pollution. 2002;**133**:349-361. DOI: 10.1023/A:101292990

[20] Chikuma M, Nishimura M. Selective sorption of fluoride ions by anion exchange resin modified with alizarin fluorine blue-praseodymium complex. Reactive Polymers. 1990;**13**:131-138. DOI: 10.1016/0923-1137(90)90047-8

[21] Puri BK, Balani S. Trace determination of fluoride using lanthanum hydroxide supported on alumina. Journal of Environmental Science and Health, Part A. 2000;**35**:109-121. DOI: 10.1080/10934.520009376957

[22] Mohan SV, Ramanaiah SV, Rajkumar B, Sharma PN. Biosorption of fluoride from aqueous solution onto algal *Spirogyra* IO1 and evaluation of adsorption kinetics. Bioresource Technology. 2007;**98**:1006-1011. DOI: 10.1016/j.biortech.2006.04.009

[23] Kemer B, Ozdes D, Gundogdu A, Bulut VN, Duran C, Soylak M. Removal of fluoride ions from aqueous solution by waste mud. Journal of Hazardous Materials. 2009;**168**:888-894. DOI: 10.1016/j.jhazmat.2009.02.109

[24] Wang SG, Ma Y, Shi YJ, Gong WX. Defluoridation performance and mechanism of nanoscale aluminum oxide hydroxide in aqueous solution. Journal of Chemical Technology and Biotechnology. 2009;**84**:1043-1050. DOI: 10.1002/jctb.2131

[25] Sundaram CS, Viswanathan N, Meenakshi S. Defluoridation of water using magnesia/chitosan composite. Journal of Hazardous Materials. 2009;**163**:618-624. DOI: 10.1016/j.jhazmat.2008.07.009

[26] Gupta VK, Ali I, Saini VK. Defluoridation of wastewaters using carbon slurry. Water Research. 2007;**41**:3307-3316. DOI: 10.1016/j.waters.2007.04.029

[27] Tor A. Removal of fluoride from an aqueous solution by using montmorillonite. Desalination. 2006;**201**:267-276. DOI: 10.1016/j.desal.2006.06.003

Chapter 8

Bioremediation of Heavy Metals

Medhat Rehan and Abdullah S. Alsohim

Abstract

Exposure to lead (Pb), zinc (Zn), cadmium (Cd), copper (Cu), and selenite ($SeO3^{-2}$) consider the main heavy metals that threat human health. These heavy metals can interfere with the function of vital cellular components. Soil heavy metal contamination represents risks to humans and the ecosystem through drinking of contaminated groundwater, direct ingestion or the food chain, and reduction in food quality. Bioremediation means cleanup of polluted environment via transformation of toxic heavy metals into less toxic form by microbes or its enzymes. Otherwise, bioremediation by microbes has limitations like production of toxic metabolites. The efflux of metal ions outside the cell, biosorption to the cell walls and entrapment in extracellular capsules, precipitation, and reduction of the heavy metal ions to a less toxic state are mechanisms to metals' resistance.

Keywords: heavy metals, bioremediation, copper, lead, cadmium, selenite

1. Introduction

Since the industrial revolution, heavy metals' waste has increased rapidly. Toxic metals' species are mobilized from industrial activities and fossil fuel consumption and eventually are accumulated through the food chain, leading to both ecological and health problems. Some of these metals are taken up as essential nutrients since they are incorporated into enzymes and cofactors. Some heavy metals exert toxic effects on microbial cells (i.e., mercury, lead, cadmium, arsenic, and silver). Mostly, resistance systems have been found on plasmids, whereas bacterial chromosomes contain genes for resistance to many of the same heavy metals' cations and oxyanions as do plasmids [1, 2]. To survive under metal-stressed conditions, bacteria have evolved several types of mechanisms to tolerate the uptake of heavy metal ions. These mechanisms include biosorption to the cell walls and entrapment in extracellular capsules, precipitation, the efflux of metal ions outside the cell, reduction of heavy metal ions to a less toxic state accumulation, and complexation of metal ions inside the cell [3, 4].

2. Copper bioremediation

In almost all life forms, copper is a metal essential for the normal function. It acts as a cofactor for a number of enzymes involved in respiration and electron transport proteins in plants, animals, and microorganisms. Copper is toxic to cells at high concentrations mainly due to the disruption of the integrity of cell membranes, its interaction with nucleic acids, interference with the energy transport system, and disruption of enzyme active sites [5–8] . At high cytoplasmic concentrations, copper can compete with other metals for their binding sites in proteins that can lead to dysfunctional proteins. Otherwise, the presence of Cu (I) in cells will react with hydrogen peroxide and produce hydroxyl radicals that will damage DNA, lipids, and other molecules [9, 10]. Resistance to copper in microorganisms is dependent mainly on three different systems:

1. The periplasmic plasmid-borne copper (pco) resistance system that encodes for PcoA, a multi-copper oxidase protein responsible for oxidation of Cu(I) in the periplasmic space. This system presents only on plasmids and presents high copper resistance [11–13].

2. The efflux ATPase pump CopA able to throw copper ions outside [10, 14].

3. Cus system (copper sensing copper efflux system) belonging to the resistance-nodulation-cell division (RND) family responsible for heavy metal export (HME-RND) that encodes especially for the CusA protein [10, 13, 15].

In agriculture, copper bactericide is considered one of the most important components in environmental contamination with copper especially in programs practiced worldwide in growing areas with citrus [16]. Many species of plant pathogenic bacteria such as *Xanthomonas citri* subsp. *citri* (Xcc) have developed resistance to copper as a consequence of using copper bactericides [5]. Copper resistance genes have taken place from strains of *X. alfalfae* subsp. *citrumelonis* from Florida and *Xanthomonas citri* subsp. *citri* from Argentina [17].

In both *X. citri* subsp. *citri* and *X. alfalfae* subsp. *citrumelonis*, the long-term use of copper bactericides has led to the development of copper-resistant (Cur) strains. In *X. citri* subsp. *citri* A44, open reading frames (ORFs) related to the genes *copL*, *copA*, *copB*, *copM*, *copG*, *copC*, *copD*, and *copF* were characterized to be present on a large (~300 kb) conjugative plasmid. The same ORFs, except *copC* and *copD*, were also present in *X. alfalfae* subsp. *citrumelonis* 1381 [5, 18]. Via molecular tools, the abundance of the copper resistance genes *cusA* and *copA*, encoding, respectively, for a Resistance Cell Nodulation protein and for a P-type ATPase pump, were assessed in Chilean marine sediment cores since, in the impacted sediment, *copA* gene was more abundant than *cusA* gene [10]. When *Sulfolobus metallicus* cells are exposed to 100 mM Cu, proteomic analysis showed that 18 out of 30 upregulated proteins are related to stress responses, the production and conversion of energy, and amino acid biosynthesis [19]. Furthermore, when searching the genome, two complete cop gene clusters encoding a Cu-exporting ATPase (CopA), metallochaperone (CopM), and a transcriptional regulator (CopT) were detected.

Based on a plate assay, Frankia strains EuI1c, CN3, QA3, and DC12 are tolerant to high levels of copper (MIC values >5 mM), while many other strains tested are very sensitive exhibiting MIC values <0.1 mM [20]. Otherwise, a 24-well growth assay was used to reexamine copper sensitivity of five *Frankia* strains. *Frankia* strains EuI1c, CN3, and DC12 showed similar growth patterns. Growth was initially inhibited at low copper concentrations (0.1 mM), but growth yield increased with elevated copper levels reaching a peak at 5 mM (**Figure 1**).

The cells grown in the elevated copper levels appeared blue which suggest that copper was accumulating inside of *Frankia* or binding to the cell surface [21]. When observed under phase-contrast microscopy, Cu^{+2}-resistant *Frankia* EuI1c formed unusual globular structures that were associated with their hyphae [20]. These structures were further investigated at a higher resolution. The increase resolution revealed that the globular structures are composed of aggregates (> 50 um) containing many smaller structures (**Figure 2**).

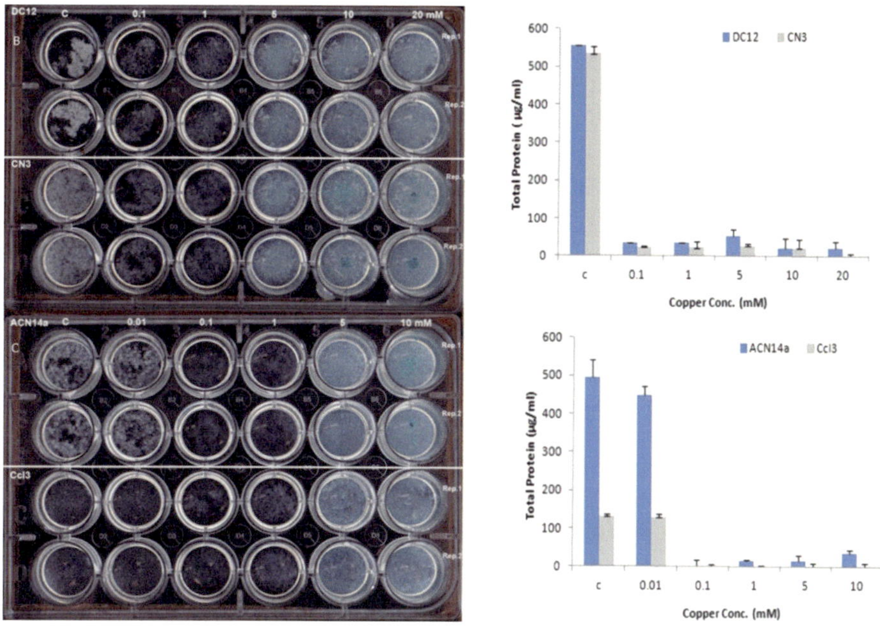

Figure 1.
Frankia strains DC12 and CN3 growth in 24-well growth system and protein assay under variant concentrations from copper stress.

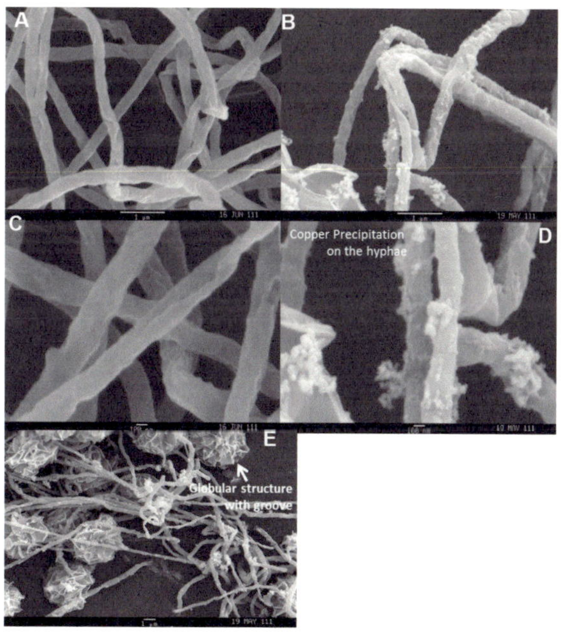

Figure 2.
SEM of Frankia strain EuI1c grown under 1 mM from copper for 1 week. Panels (A and C) Control conditions against (B, D and E) copper metal condition. Size bars represent: 1 µm.

These smaller structures were about 5 µm in diameter and were also observed as individual structures throughout the hyphae. At higher magnification, the structures have a grooved pattern and appear connected to the hyphae by amorphous material [21]. Similar globular structures were observed with SEM of other copper-resistant

Frankia strains (e.g., strain DC12). These observations suggest that *Frankia* may precipitate the Cu^{+2}-phosphate complex to the hyphae. *Acidithiobacillus ferrooxidans* will detoxify Cu^{+1} metal by formatting phosphate granules through stimulation of polyphosphate hydrolysis and formation of metal-phosphate complexes [22].

The elemental composition analysis of these structures was investigated by the use of SEM-EDAX. As expected, these structures exhibited an elevated copper content that was represented by a 73-fold more than the control increase in the intensity but also contained an elevated phosphate content that was about 43.88-fold higher intensity level than the control cells. Furthermore, the oxygen content increased 3.5-fold under copper-stressed condition. All three of these elements had nearly the same intensity values under Cu^{+2} condition. These results suggest that a copper-phosphate compound forms and binds to *Frankia* cell surface. The EDAX spectra showed that the bodies present in the cells were mainly composed of phosphorus and oxygen [8, 21]. The highly sensitive MS analysis of excised bands produced peptides such as periplasmic binding protein/LacI transcriptional regulator (E3IXA6; FraEuI1c_7040 gene) with the appropriate protein size (37.6 kDa). Another protein of interest was the sulfate ABC transporter, a periplasmic sulfate-binding protein (E3J029; FraEuI1c_1092; 36.6 kDa) which had 2, 5, and 20 peptides. These data would indicate a tenfold increase in expression under 2 mM Cu^{+2}-stress, while the extracellular ligand-binding receptor (39.925 kDa) induced up to six- and eightfold under 1 and 2 mM copper, respectively. These proteins may be playing a role in copper resistance through binding and accumulating copper as in the periplasmic binding protein/LacI transcriptional regulator and extracellular ligand-binding receptor or transporting copper outside the cell as in sulfate ABC transporter, periplasmic sulfate-binding receptor. The relative expression of the heavy metal transporter/detoxification gene (FraEuI1c_6308) and copper-translocating P-type ATPase (FraEuI1c_6307) has shown 30- to 35-fold increase in the level of expression compared to the control under Cu^{+2}-stress for 8 days. These results suggest that these two gene products may play a role in copper tolerance [21].

In some bacteria and algae, it has been proposed that inorganic polyphosphates and transport of metal-phosphate complexes will participate in heavy metal tolerance [23]. After the *Frankia* grew under copper condition, the level of phosphate in EDAX analysis was high which would support this hypothesis of the formation of a metal-phosphate complex. This complex could be effluxed outside the cell via P-type ATPase or phosphate efflux system [22]. In *Enterococcus hirae*, CopA functions to import copper when it is deficient [24]. With *Pseudomonas syringae*, CopA is an outer membrane protein and functions in the sequestration and compartmentalization of copper in the periplasm and outer membrane [25]. The function of the CopB protein in *E. hirae* is to remove excess copper present in the cytoplasm [24]. The specific function of CopB protein in ***E. coli*** and *Pseudomonas syringae* is not yet defined [26]. With *E. hirae*, *copA* and *copB* are involved in copper transport using ATPases, while *copY* gene product acts as a copper-responsive repressor. The *copZ* functions in the transport of intracellular copper.

3. Lead bioremediation

Lead enters the cells through Fe^{2+} and Ca^{2+} transporters and then exerts its toxicity by displacing these cations at their binding sites in metalloproteins. Heavy metal resistance systems in many bacterial are based on efflux. Two groups of efflux systems have been recognized in gram-negative bacteria which are chemiosmotic pumps, e.g., the three-component divalent-cation efflux systems of *Ralstonia metallidurans* (*cnr*, *ncc*, and *czc* [27] and/or P-type ATPases, e.g., the Zn(II), Cu(II), and Cd(II) ATPases [28, 29]. In both gram-negative and gram-positive bacteria, lead resistance has been reported in lead-contaminated soils. ***Bacillus megaterium*** demonstrating intracellular cytoplasmic leads to accumulation and *Pseudomonas marginalis* showing extracellular leads to exclusion [30]. Furthermore, the ***Staphylococcus aureus*** and *Citrobacter freundii* accumulated the metal as an intracellular lead-phosphate [31]. CadA ATPase of ***Staphylococcus aureus*** and the ZntA ATPase of ***Escherichia coli*** have been reported as efflux of Pb(II) [32].

Furthermore, 27 isolates were isolated from some abandoned mining areas in Morocco and found to belong to *Streptomyces* and Amycolatopsis genera. The minimum inhibitory concentration (MIC) recorded was 0.1 $mg{\cdot}mL^{-1}$ for both Zn and Cu, 0.55 for Pb, and 0.15 for Cr. Chemical precipitation assay revealed that the 27 isolates have a strong ability to accumulate Pb (up to 600 mg of Pb/g of biomass for *Streptomyces* sp. BN3) [33].

Interplay between CBA transporters and P-type ATPases in *Cupriavidus metallidurans* CH34 for zinc and cadmium resistance is reported [34]. The pbrTRABCD gene cluster from *Cupriavidus metallidurans* CH34 revealed that export of Zn^{2+}, Cd^{2+} and Pb^{2+} was via the main transporter component of the operon P-type ATPase PbrA, whereas PbrB, the second component of the operon, was shown to be a phosphatase that increased lead resistance. P-type ATPase that removes Pb^{2+} ions from the cytoplasm and a phosphatase that produces inorganic phosphate for lead sequestration in the periplasm represent the new lead resistance model in *Cupriavidus metallidurans* CH34. In several different bacterial species and when searching databases, gene clusters containing neighboring genes for P-type ATPase and phosphatase were detected which suggest that Pb^{2+} detoxification via active efflux and sequestration may be a widespread mechanism of resistance [34]. In *Pseudomonas putida* KT2440, two P-type ATPases and two CBA transporters exhibited that resistance mechanisms for Zn^{2+} and Cd^{2+} are somewhat different than for Pb^{2+} since Zn^{2+} and Cd^{2+} cannot be sequestered as insoluble compounds easily [32, 34].

A group of transporters, the cation diffusion facilitator family (CDF), can catalyze heavy metal influx or efflux in both prokaryotes and eukaryotes. All characterized CDF proteins to date can transport metals only (such as Fe^{2+}, Co^{2+}, Ni^{2+}, Zn^{2+} and Cd^{2+}), in contrast to other protein families, such as P-type ATPases or CBA transporters. In *C. metallidurans*, CDF family of chemiosmotic efflux systems with the CzcD Cd^{2+} and Zn^{2+} efflux system was first described [34]. CDF transporters provide very low resistance level, but it plays a main role in heavy metal buffer at low concentration of the metal in the cell cytoplasmic [34].

Detoxification mechanism for Pb^{2+} can also be achieved by sequestration. In several bacterial species and via the use of intra- and extracellular binding of Pb^{2+}, they can avoid toxicity as in ***S. aureus***, *Citrobacter freundii* [35, 31], and *Vibrio harveyi* [36] by precipitating lead as a phosphate salt. Mainly through exopolysaccharides (EPSs), binding of heavy metals can take place. EPS could act as a biosorbent of free metal ions, but it cannot be considered as inducible resistance mechanism in response to metals [34].

Nine candidate core biomarker genes might be tightly correlated with the response or transport of heavy metals. These genes, namely, *NILR1*, *PGPS1*, *WRKY33*, *BCS1*, *AR781*, *CYP81D8*, *NR1*, *EAP1*, and *MYB15*. The same expression trend and response to different stresses (Cd, Pb, and Cu) by experimental results have been shown [37].

The mechanism of lead resistance in *Frankia* sp. strain EAN1pec has been reported which include cells' accumulated Pb^{2+} with saturation kinetics (**Figure 3**). The Cu^{2+}-ATPase and cation diffusion facilitator (CDF) in addition to several hypothetical transporters were upregulated under lead stress that may indicate

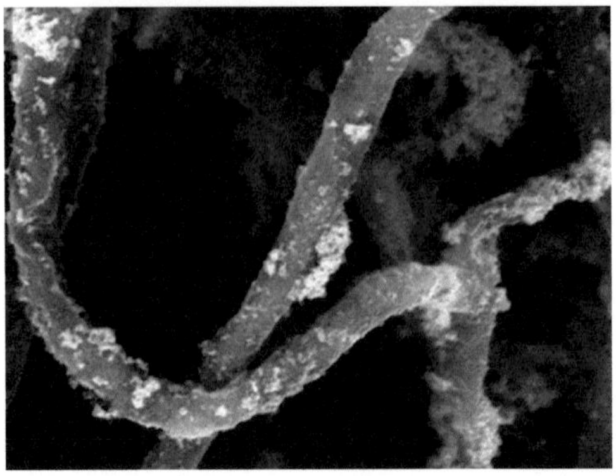

Figure 3.
Lead precipitation on Frankia hyphae.

metal export. Furthermore, a potential transcription factor (DUF156) binding site associated with several proteins was identified with heavy metals [38]. The EDAX results showed much high proportion of phosphate in *Frankia* cultures exposed to higher Pb^{2+} concentrations which could indicate different Pbx(PO4)x compounds formed, which bind to *Frankia* cell surface [38].

4. Cadmium bioremediation

Cadmium (Cd^{2+}), the heavy metal, is toxic in its ionized form to microbes and humans. It is found in the biosphere and often associated with zinc ores at concentrations approaching 0.01–1.8 ppm. It can enter the bacterial cell normally by essential divalent cations via transport systems. Cadmium toxicity has effect by inhibiting respiration via binding to essential proteins' sulfhydryl groups and can also cause single-strand breakage of DNA in ***E. coli*** [39].

The full resistance to Cd^{2+} required the interplay of a P-type ATPase that exported cytoplasmic ions to periplasm and a CBA transporter that further exported periplasmic ions to the outside. Furthermore, membrane transport pumps export metal ions from the cell and binding factors involved in creating tolerance to heavy metal ions through detoxify metals by sequestration (i.e., cell wall components (exopolysaccharides) and intracellular binding proteins (like metallothioneins and metallochaperones)) [34]. As cytoplasmic metal cation-binding proteins, metallothioneins can lower the concentrations of free ion in the cytoplasm. SmtA from *Synechococcus* PCC 7942 was the first metallothionein characterized in bacteria and can sequester and detoxify Zn^{2+} and Cd^{2+}. Otherwise, SmtB is a repressor which can dissociate from DNA in the presence of metals [40–43].

In *Streptococcus thermophilus* Strain 4134, two genes (*cadCSt* and *cadASt*) were confirmed to constitute in cadmium/zinc resistance. P-type cadmium efflux ATPases are the proposed product of the *cadA* open reading frame (CadA*St*), whereas ArsR-type regulatory proteins are the predicted proteins encoded by *cadCSt* (CadC*St*) [39]. The plasmid-encoded *cad* system in ***S. aureus*** is the best characterized Cd(II) resistance efflux system. CadA functions as an efflux pump that exports Cd(II) from the cell interior [44–46]. The gene product of *cadC* binds Cd(II) as it proposed inside the organism since cadC can bind two Cd(II) ions via a pair of cysteine residues. It is proposed that cadA takes Cd(II) from cadC in the cytoplasmic membrane [47]. *cadD*, the cadmium resistance gene, has been identified in a two-component operon which contains the resistance gene *cadD* and an inactive regulatory gene, *cadX*, from the *Staphylococcus aureus* plasmid pRW001

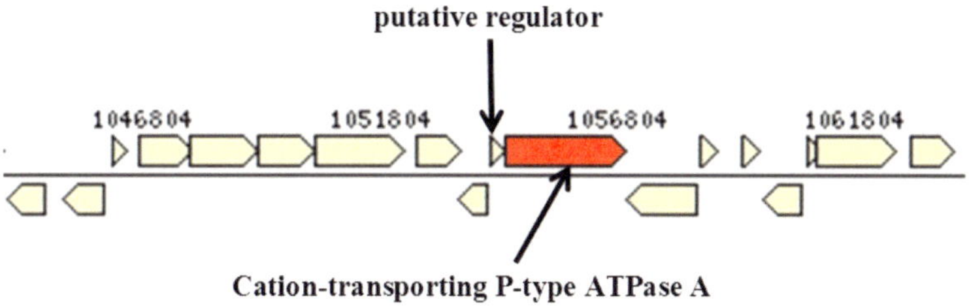

Figure 4.
The proposed CadA gene in Frankia alni ACN14a.

[48]. ZntA, the metal-dependent ATP hydrolysis activity which exports Cd(II), Pb(II), and Zn(II) from ***Escherichia coli***, is a cation-translocating ATPase. ZntA expression is mediated by transcriptional regulator protein ZntR, belonging to the MerR family. Based on in vitro molecular cloning analysis and in silico studies, P_{cadR} and CadR are active in the presence of Cd with the highest binding affinity between the CadR protein and P_{cadR} [45, 49].

The *Frankia* strain ACN14a and EuI1c genome was first searched for cadmium-binding motifs in COG and Pfam databases. A BLASTP analysis was performed on the *Frankia* ACN14a and EuI1c genome using the known CadA proteins as a query sequence. Blasting the published *Frankia* genomes against functionally identified CadA and putative cobalt-zinc-cadmium resistance amino acid sequences has revealed two possible genes (FRAAL0989 and FRAAL3628).

The identified gene which is *CadA* (FRAAL0989) in *Frankia* ACN14a is predicted to encode cation-transporting P-type ATPase A that possess cadmium and zinc outside the cells (**Figure 4**). Moreover, the putative cobalt-zinc-cadmium resistance (FRAAL3628) in the same strain is expected to work as transmembrane protein and consider cobalt-zinc-cadmium efflux system protein.

5. Selenite reduction

Selenium, in the form of selenocysteine or selenomethionine, is found in several stress proteins including glutathione peroxidase, alkyl hydroperoxidase, and multiple disulfide reductases. The deprotonated electrons of selenium cofactors make the selenoproteins' reduction–oxidation reactive, explaining why many identified selenoproteins are involved in thiol and oxidative stress resistance. Since selenite generates these stresses in the cell, the stress-related selenoproteins may function doubly in detoxification and removal of free selenite ions from the cytoplasm. About 20% of sequenced bacteria contain selenoproteins [50].

The detoxification mechanism of selenite reduction in aerobic condition by microorganisms is not yet fully elucidated. Previously, it has been reported that selenite reduction may be catalyzed by a periplasmic nitrate reductase as in a selenate reductase, a periplasmic nitrate reductase in *Thauera selenatis* [51], a molybdenum-dependent membrane-bound enzyme of ***Enterobacter cloacae*** SLD1a-1 [52], *Thiosphaera pantotropha* [53], a periplasmic cytochrome B in *Thauera selenatis* [54], or a hydrogenase 1 of *Clostridium pasteurianum* [55]. Recent studies have indicated that NADPH-/NADH-dependent selenate reductase enzymes bring about the reduction of selenium (selenite/selenate) oxyanions. Selenite can be reduced to inert elemental selenium, which occurs in the selenite-resistant *Frankia* strains CN3, EuI1c, EUN1f, and DC12 [20].

However, all of the *Frankia* genomes contained synthase proteins for small thiols like mycothiol (MSH), which may substitute for glutathione for metal resistance.

Selenium can also be reduced enzymatically either using thioredoxin and its reductase (TrxA and TrxB) or other oxyanion reductases, whereas selenite reduction by fumarate reductase (FccA) in the periplasm was identified in *Shewanella oneidensis* MR-1 [56]. *Frankia* CN3 has a second type of nitrate reductase (NasC) located with the nitrite reductases (NirBD) that may also contribute to its greater selenite resistance.

Many bacteria including ***Enterobacter cloacae*** SLD1a-1, ***Bacillus megaterium***, *Comamonas testosteroni* S44, *Thauera selenatis*, *Rhodopseudomonas palustris* Strain N, and ***Bacillus cereus*** form nanospheres of Se° during selenite or selenate reduction [57–63], whereas under aerobic conditions, ***B. cereus*** reduces selenite to Se° nanospheres in the size range of 150–200 nm [64]. *Rhodospirillum rubrum* is postulated to efficiently transport elemental selenium out of the cell [65]. This hypothesis is supported by the results of ultracentrifugation experiments showing that the buoyant density of cells increases in the presence of selenite during the reduction phase. With ***Desulfovibrio desulfuricans***, selenium-containing particles are postulated to be formed in the cytoplasm. However, the red elemental selenium that accumulates in the media during the stationary growth phase is released, the result of cell lysis. On the surfaces of ***E. cloacae*** cells grown in the presence of selenite, more or less spherical protrusions were observed [66]. Selenium-containing particles were observed in the culture medium, but intracellular Se° was not detected in this study. Selenite reduction was suggested to occur via a membrane-associated reductase that was followed by rapid expulsion of the Se particles. Our data shows extra- and intracellular nanosphere particles which may be transported through the membrane. My hypothesis is that the small particles are transported out of the cell and then form the large particles observed in the culture medium by extracellular aggregation. This postulated mechanism of transport would require an extremely large amount of energy.

In summary, selenite resistance may result from oxidation of selenite to the less toxic selenate using SorA. *Frankia* selenite resistance is likely due to alternate sulfate transporters (CysPUWA) that prevent sulfur starvation. The selenite reduction observed in resistant strains could occur through several mechanisms including NasC/NirBD or mycothiol, TrxAB, and YedY.

Frankia strain EuI1c showed a pattern of resistance to selenite. Growth steadily decreased as selenite levels elevated reaching a plateau at 3 mM that remained constant up to 8 mM. Strain EuI1c showed a MTC value of <0.1 mM, while the MIC was 3 mM. Strain CN3 showed a different overall pattern and a modest level of selenite resistance. This strain was more sensitive to 0.1 mM levels than higher levels (1–5 mM). Both of these strains formed a reddish cell suspension in the 24-well plates. These results indicate the reduction of the toxic, soluble, and colorless sodium selenite (Na_2SeO3^{2-}) to the nontoxic, insoluble, and red-colored elemental selenium form (Se°). The red color development started to appear in these cultures after the 48-h incubation [67]. Visual observation of the cultures implies that *Frankia* will reduce colorless selenite to red-colored elemental selenium, which is nontoxic and insoluble [60]. Cells exposed to 0.1 mM selenite reduced completely all of the selenite at day 5. At 0.5 mM selenite, the Se° production initiated at day 3 and ended at day 8. However, at 1 mM culture, Se° production initiated at day 3 and reached saturation at day 7.

When examined under scanning electron microscope, selenite-resistant *Frankia* EuI1c formed spherical nanospheres (**Figure 5**). These nanospheres were associated with the hyphae outside the cell as free deposits and also appeared as aggregates attached to the hyphae mass. Furthermore, these spherical particles also appear to be located inside the hyphae. The mentioned nanospheres may composed from reduced formed Se°. Since the reduction may occur in the cytoplasm, the nanoparticles would be exported outside. These nanospheres were observed in different sizes in the nanometer range [67, 68].

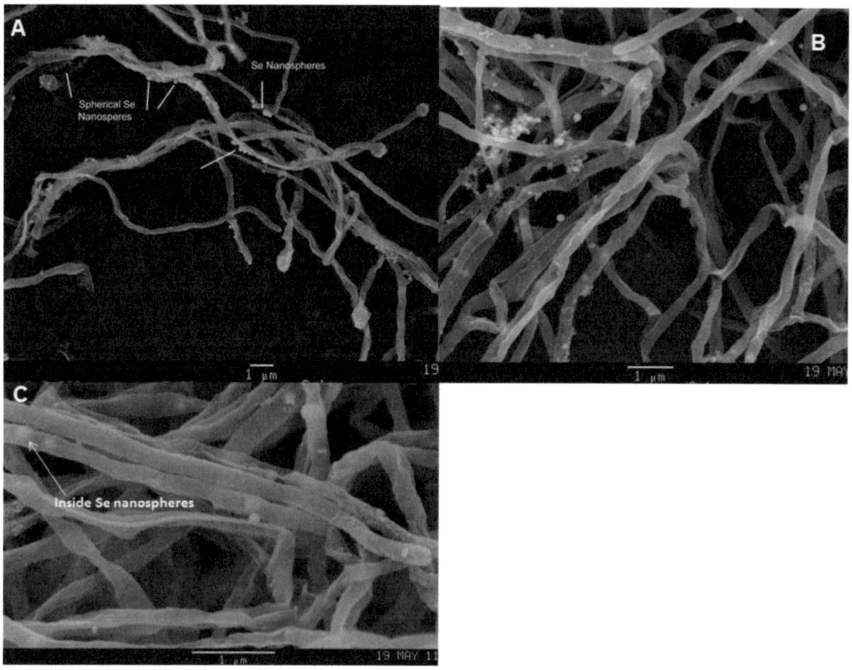

Figure 5.
Se nanospheres formed inside and outside Frankia hyphae. Panels (A, B and C) represent selenite oxyanions condition. Size bars represent: 1 μm.

The elemental composition analysis of these nanospheres was investigated by the use of SEM-EDAX. As predicted, these nanospheres exhibited an elevation. Three absorption peaks in EDAX analysis at 1.37 keV (peak SeLα), 11.22 keV (peak SeKα), and 12.49 keV (peak SeKß) can be produced from selenium absorption. The first peak is related to 1.37 keV (peak SeLα) (keV = kilo electron Volt), whereas the second peak is met with 11.22 keV (peak SeKα) [61, 64].

6. Conclusion

Heavy metals are harmful to human health via interference with the function of vital cellular components. Lead (Pb), cadmium (Cd), copper (Cu), and selenite ($SeO3^{2-}$) are metals and metalloids that are widespread in the environment. P-type ATPase system that exported cytoplasmic ions to the periplasm and a CBA transporter that further exported periplasmic ions to the outside are general mechanisms in resistance Co, Pb, and Cd. Furthermore, in metals detoxification by sequestration, binding factors will be involved in creating tolerance to heavy metal ions.

Acknowledgements

We thank Teal Furnholm and Robert Mooncy for their help with the photography and Nancy Chemin for her help with the electron microscopy.

Conflict of interest

The authors declare no conflict of interest.

Author details

Medhat Rehan[1,2*] and Abdullah S. Alsohim[2]

1 Department of Genetics, Kafrelsheikh University, Kafr El-Sheikh, Egypt

2 Department of Plant Production and Protection, College of Agriculture and Veterinary Medicine, Qassim University, Saudi Arabia

*Address all correspondence to: medhat.rehan@agr.kfs.edu.eg; m.rehan@qu.edu.sa

References

[1] Igiri BE, Okoduwa SIR, Idoko GO, Akabuogu EP, Adeyi AO, Ejiogu IK. Toxicity and bioremediation of heavy metals contaminated ecosystem from tannery wastewater: A Review. Journal of Toxicology. 2018;**2018**:16

[2] Mohamed RM, Abo-Amer AE. Isolation and characterization of heavy-metal resistant microbes from roadside soil and phylloplane. Journal of Basic Microbiology. 2012;**52**(1):53-65

[3] Nies DH. Efflux-mediated heavy metal resistance in prokaryotes. FEMS Microbiology Reviews. 2003;**27**(2-3):313-339

[4] Nies DH. Microbial heavy-metal resistance. Applied Microbiology and Biotechnology. 1999;**51**(6):730-750

[5] Behlau F, Canteros BI, Jones JB, Graham JH. Copper resistance genes from different xanthomonads and citrus epiphytic bacteria confer resistance to Xanthomonas citri subsp. citri. European Journal of Plant Pathology. 2012;**133**(4):949-963

[6] García-Horsman JA, Barquera B, Rumbley J, Ma J, Gennis RB. The superfamily of heme-copper respiratory oxidases. Journal of Bacteriology. 1994;**176**(18):5587-5600

[7] Cervantes C, Gutierrez-Corona F. Copper resistance mechanisms in bacteria and fungi. FEMS Microbiology Reviews. 1994;**14**(2):121-137

[8] Rehan M. Microbial bioremediation: A Review. Journal of Agricultural and Veterinary Sciences. 2017;**10**(2):147-162

[9] Harrison MD, Jones CE, Solioz M, Dameron CT. Intracellular copper routing: The role of copper chaperones. Trends in Biochemical Sciences. 2000;**25**(1):29-32

[10] Besaury L, Bodilis J, Delgas F, Andrade S, De la Iglesia R, Ouddane B, et al. Abundance and diversity of copper resistance genes cusA and copA in microbial communities in relation to the impact of copper on Chilean marine sediments. Marine Pollution Bulletin. 2013;**67**(1-2):16-25

[11] Brown NL, Barrett SR, Camakaris J, Lee BT, Rouch DA. Molecular genetics and transport analysis of the copper-resistance determinant (pco) from Escherichia coli plasmid pRJ1004. Molecular Microbiology. 1995;**17**(6):1153-1166

[12] Gittins JR. Cloning of a copper resistance gene cluster from the cyanobacterium Synechocystis sp. PCC 6803 by recombineering recovery. FEBS Letters. 2015;**589**(15):1872-1878

[13] Staehlin BM, Gibbons JG, Rokas A, O'Halloran TV, Slot JC. Evolution of a heavy metal homeostasis/Resistance Island reflects increasing copper stress in Enterobacteria. Genome Biology and Evolution. 2016;**8**(3):811-826

[14] Rensing C, Fan B, Sharma R, Mitra B, Rosen BP. CopA: An Escherichia coli Cu(I)-translocating P type ATPase. Proceedings of the National Academy of Sciences of the United States of America. 2000;**97**(2):652-656

[15] Piddock LJ. Multidrug-resistance efflux pumps—Not just for resistance. Nature Reviews. Microbiology. 2006;**4**(8):629-636

[16] Leite RP, Mohan SK. Integrated management of the citrus bacterial canker disease caused by Xanthomonas campestris pv. Citri in the state of Paraná, Brazil. Crop Protection. 1990;**9**(1):3-7

[17] Behlau F, Hong JC, Jones JB, Graham JH. Evidence for acquisition of copper resistance genes from

different sources in citrus-associated xanthomonads. Phytopathology. 2013;**103**(5):409-418

[18] Behlau F, Canteros BI, Minsavage GV, Jones JB, Graham JH. Molecular characterization of copper resistance genes from Xanthomonas citri subsp. citri and Xanthomonas alfalfae subsp. citrumelonis. Applied and Environmental Microbiology. 2011;**77**(12):4089-4096

[19] Orell A, Remonsellez F, Arancibia R, Jerez CA. Molecular characterization of copper and cadmium resistance determinants in the biomining Thermoacidophilic archaeon *Sulfolobus metallicus*. Archaea. 2013;**2013**:16

[20] Richards JW, Krumholz GD, Chval MS, Tisa LS. Heavy metal resistance patterns of Frankia strains. Applied and Environmental Microbiology. 2002;**68**(2):923-927

[21] Rehan M, Furnholm T, Finethy RH, Chu F, El-Fadly G, Tisa LS. Copper tolerance in Frankia sp. strain EuI1c involves surface binding and copper transport. Applied Microbiology and Biotechnology. 2014;**98**(18):8005-8015

[22] Alvarez S, Jerez CA. Copper ions stimulate polyphosphate degradation and phosphate efflux in Acidithiobacillus ferrooxidans. Applied and Environmental Microbiology. 2004;**70**(9):5177-5182

[23] Remonsellez F, Orell A, Jerez CA. Copper tolerance of the thermoacidophilic archaeon *Sulfolobus metallicus*: Possible role of polyphosphate metabolism. Microbiology. 2006;**152**(Pt 1):59-66

[24] Solioz M, Stoyanov JV. Copper homeostasis in enterococcus hirae. FEMS Microbiology Reviews. 2003;**27**(2-3):183-195

[25] Mellano MA, Cooksey DA. Nucleotide sequence and organization of copper resistance genes from pseudomonas syringae pv. Tomato. Journal of Bacteriology. 1988;**170**(6):2879-2883

[26] Masami N, Masao G, Katsumi A, Tadaaki H. Nucleotide sequence and organization of copper resistance genes from pseudomonas syringae pv. Actinidiae. European Journal of Plant Pathology. 2004;**110**(2):223-226

[27] Taghavi S, Mergeay M, Nies D, van der Lelie D. Alcaligenes eutrophus as a model system for bacterial interactions with heavy metals in the environment. Research in Microbiology. 1997;**148**(6):536-551

[28] Silver S, Nucifora G, Phung LT. Human Menkes X-chromosome disease and the staphylococcal cadmium-resistance ATPase: A remarkable similarity in protein sequences. Molecular Microbiology. 1993;**10**(1):7-12

[29] Borremans B, Hobman JL, Provoost A, Brown NL, van Der Lelie D. Cloning and functional analysis of the pbr lead resistance determinant of Ralstonia metallidurans CH34. Journal of Bacteriology. 2001;**183**(19):5651-5658

[30] Roane TM. Lead resistance in two bacterial isolates from heavy metal-contaminated soils. Microbial Ecology. 1999;**37**(3):218-224

[31] Levinson HS, Mahler I. Phosphatase activity and lead resistance in Citrobacter freundii and Staphylococcus aureus. FEMS Microbiology Letters. 1998;**161**(1):135-138

[32] Rensing C, Sun Y, Mitra B, Rosen BP. Pb(II)-translocating P-type ATPases. The Journal of Biological Chemistry. 1998;**273**(49):32614-32617

[33] El Baz S, Baz M, Barakate M, Hassani L, El Gharmali A, Imziln B, et al. Resistance to and accumulation of

heavy metals by Actinobacteria isolated from abandoned mining areas. The Scientific World Journal. 2015;**2015**:14

[34] Hynninen A. Zinc, cadmium and lead resistance mechanisms in bacteria and their contribution to biosensing. 2010

[35] Levinson HS, Mahler I, Blackwelder P, Hood T. Lead resistance and sensitivity in Staphylococcus aureus. FEMS Microbiology Letters. 1996;**145**(3):421-425

[36] Mire CE, Tourjee JA, O'Brien WF, Ramanujachary KV, Hecht GB. Lead precipitation by Vibrio harveyi: Evidence for novel quorum-sensing interactions. Applied and Environmental Microbiology. 2004;**70**(2):855-864

[37] Niu C, Jiang M, Li N, Cao J, Hou M, Ni D-a, et al. Integrated bioinformatics analysis of As, Au, Cd, Pb and Cu heavy metal responsive marker genes through Arabidopsis thaliana GEO datasets. PeerJ. 2019;7:e6495

[38] Furnholm T, Rehan M, Wishart J, Tisa LS. Pb2+ tolerance by Frankia sp. strain EAN1pec involves surface-binding. Microbiology. 2017;**163**(4):472-487

[39] Schirawski J, Hagens W, Fitzgerald GF, Van Sinderen D. Molecular characterization of cadmium resistance in Streptococcus thermophilus strain 4134: An example of lateral gene transfer. Applied and Environmental Microbiology. 2002;**68**(11):5508-5516

[40] Olafson RW, McCubbin WD, Kay CM. Primary- and secondary-structural analysis of a unique prokaryotic metallothionein from a Synechococcus sp. cyanobacterium. The Biochemical Journal. 1988;**251**(3):691-699

[41] Turner JS, Morby AP, Whitton BA, Gupta A, Robinson NJ. Construction of Zn2+/Cd2+ hypersensitive cyanobacterial mutants lacking a functional metallothionein locus. The Journal of Biological Chemistry. 1993;**268**(6):4494-4498

[42] Turner JS, Glands PD, Samson AC, Robinson NJ. Zn2+−sensing by the cyanobacterial metallothionein repressor SmtB: Different motifs mediate metal-induced protein-DNA dissociation. Nucleic Acids Research. 1996;**24**(19):3714-3721

[43] Naz N, Young HK, Ahmed N, Gadd GM. Cadmium accumulation and DNA homology with metal resistance genes in sulfate-reducing bacteria. Applied and Environmental Microbiology. 2005;**71**(8):4610-4618

[44] Oger C, Berthe T, Quillet L, Barray S, Chiffoleau JF, Petit F. Estimation of the abundance of the cadmium resistance gene cadA in microbial communities in polluted estuary water. Research in Microbiology. 2001;**152**(7):671-678

[45] Prabhakaran R, Rajkumar SN, Ramprasath T, Selvam GS. Identification of promoter PcadR, in silico characterization of cadmium resistant gene cadR and molecular cloning of promoter PcadR from Pseudomonas aeruginosa BC15. Toxicology and Industrial Health. 2018;**34**(12):819-833

[46] Oger C, Mahillon J, Petit F. Distribution and diversity of a cadmium resistance (cadA) determinant and occurrence of IS257 insertion sequences in staphylococcal bacteria isolated from a contaminated estuary (Seine, France). FEMS Microbiology Ecology. 2003;**43**(2):173-183

[47] Bruins MR, Kapil S, Oehme FW. Microbial resistance to metals in the environment. Ecotoxicology and Environmental Safety. 2000;**45**(3):198-207

[48] Crupper S, Worrell V, Stewart G, Iandolo J. Cloning and expression of

cadD, a new cadmium resistance gene of *Staphylococcus aureus*. Journal of Bacteriology. 1999;**181**(13):4071-4075

[49] Binet MR, Poole RK. Cd(II), Pb(II) and Zn(II) ions regulate expression of the metal-transporting P-type ATPase ZntA in Escherichia coli. FEBS Letters. 2000;**473**(1):67-70

[50] Zhang Y, Gladyshev VN. General trends in trace element utilization revealed by comparative genomic analyses of Co, Cu, Mo, Ni, and Se. The Journal of Biological Chemistry. 2010;**285**(5):3393-3405

[51] Schroder I, Rech S, Krafft T, Macy JM. Purification and characterization of the selenate reductase from *Thauera selenatis*. The Journal of Biological Chemistry. 1997;**272**(38):23765-23768

[52] Watts CA, Ridley H, Condie KL, Leaver JT, Richardson DJ, Butler CS. Selenate reduction by Enterobacter cloacae SLD1a-1 is catalysed by a molybdenum-dependent membrane-bound enzyme that is distinct from the membrane-bound nitrate reductase. FEMS Microbiology Letters. 2003;**228**(2):273-279

[53] Berks BC, Richardson DJ, Robinson C, Reilly A, Aplin RT, Ferguson SJ. Purification and characterization of the periplasmic nitrate reductase from Thiosphaera pantotropha. European Journal of Biochemistry. 1994;**220**(1):117-124

[54] Krafft T, Bowen A, Theis F, Macy JM. Cloning and sequencing of the genes encoding the periplasmic-cytochrome B-containing selenate reductase of Thauera selenatis. DNA Sequence. 2000;**10**(6):365-377

[55] Yanke LJ, Bryant RD, Laishley EJ. Hydrogenase I of clostridium pasteurianum functions as a novel selenite reductase. Anaerobe. 1995;**1**(1):61-67

[56] Li D-B, Cheng Y-Y, Wu C, Li W-W, Li N, Yang Z-C, et al. Selenite reduction by Shewanella oneidensis MR-1 is mediated by fumarate reductase in periplasm. Scientific Reports. 2014;**4**:3735

[57] Javed S, Sarwar A, Tassawar M, Faisal M. Conversion of selenite to elemental selenium by indigenous bacteria isolated from polluted areas. Chemical Speciation & Bioavailability. 2015;**27**(4):162-168

[58] Kora AJ. *Bacillus cereus*, selenite-reducing bacterium from contaminated lake of an industrial area: A renewable nanofactory for the synthesis of selenium nanoparticles. Bioresources and Bioprocessing. 2018;**5**(1):30

[59] Mishra RR, Prajapati S, Das J, Dangar TK, Das N, Thatoi H. Reduction of selenite to red elemental selenium by moderately halotolerant bacillus megaterium strains isolated from Bhitarkanika mangrove soil and characterization of reduced product. Chemosphere. 2011;**84**(9):1231-1237

[60] Yee N, Ma J, Dalia A, Boonfueng T, Kobayashi DY. Se(VI) reduction and the precipitation of Se(0) by the facultative bacterium Enterobacter cloacae SLD1a-1 are regulated by FNR. Applied and Environmental Microbiology. 2007;**73**(6):1914-1920

[61] Zheng S, Su J, Wang L, Yao R, Wang D, Deng Y, et al. Selenite reduction by the obligate aerobic bacterium Comamonas testosteroni S44 isolated from a metal-contaminated soil. BMC Microbiology. 2014;**14**(1):204

[62] Debieux CM, Dridge EJ, Mueller CM, Splatt P, Paszkiewicz K, Knight I, et al. A bacterial process for selenium nanosphere assembly. Proceedings of the National Academy of Sciences. 2011;**108**(33):13480

[63] Li B, Liu N, Li Y, Jing W, Fan J, Li D, et al. Reduction of selenite

to red elemental selenium by Rhodopseudomonas palustris strain N. PLoS One. 2014;**9**(4):e95955

[64] Dhanjal S, Cameotra SS. Aerobic biogenesis of selenium nanospheres by Bacillus cereus isolated from coalmine soil. Microbial Cell Factories. 2010;**9**:52

[65] Kessi J, Ramuz M, Wehrli E, Spycher M, Bachofen R. Reduction of selenite and detoxification of elemental selenium by the phototrophic bacterium Rhodospirillum rubrum. Applied and Environmental Microbiology. 1999;**65**(11):4734-4740

[66] Losi ME, Frankenberger WT. Reduction of selenium oxyanions by Enterobacter cloacae SLD1a-1: Isolation and growth of the bacterium and its expulsion of selenium particles. Applied and Environmental Microbiology. 1997;**63**(8):3079-3084

[67] Rehan M, Alsohim AS, El-Fadly G, Tisa LS. Detoxification and reduction of selenite to elemental red selenium by Frankia. Antonie Van Leeuwenhoek. 2019;**112**(1):127-139

[68] Gonzalez-Gil G, Lens PNL, Saikaly PE. Selenite reduction by anaerobic microbial aggregates: Microbial community structure, and proteins associated to the produced selenium spheres. Frontiers in Microbiology. 2016;7:571

Chapter 9

Bioremediation of Petroleum-Contaminated Soil

Shuisen Chen and Ming Zhong

Abstract

Petroleum is not only an important energy resource to boost the economic development, but also a major pollutant of the soil. The toxicity of petroleum can cause a negative impact on ecosystem, as well as the negative effects related to its carcinogenic for both animals and humans. In the present study, bioremediation as an alternative tool for restoration petroleum-contaminated soils was set forth, and focusing on the phytoremediatior plants, petroleum-biodegradable microorganism are responsible for the biodegradation of petroleum. In the present chapter, the bioremediation of petroleum-contaminated soil, as well as the influence factors of bioremediation are elaborated based on the recently studies. This will provide a novel understanding on bioremediation and help improve strategies for petroleum-contaminated soils remediation.

Keywords: petroleum, contaminated soil, bioremediation, phytoremediation, rhizoremediation, biostimulation, bioaugmentation

1. Introduction

Petroleum is an important strategic resource that dominates the world economy [1]. Petroleum is composed of a complex mixture of aromatic hydrocarbons, aliphatic hydrocarbons, heterocyclic hydrocarbons, asphaltenes and non-hydrocarbon compounds. And 60–90% of them are classified as biodegradable [2]. In the past decades, with the development of petroleum industry, petroleum has caused a severe environment contamination and relevant adverse effects during the exploration, transportation, management or storage of hydrocarbons in underground deposits, and refining processes, among which the soil contamination with petroleum is a serious global problem [3].

The petroleum contamination induces oxidative stress, causes the alteration in soil's chemical composition and low nutrient availability. The primary harmful effects of petroleum include inhibition of seed germination, reduction of photosynthetic pigments, slowdown of nutrient assimilation, inhibition of root growth, foliar deformation and tissue necrosis, as well as destroy biological membranes, disturb the signaling of metabolic pathways and disrupt plant roots architecture [4–7]. The low-molecular-weight hydrocarbons can penetrate plant cells resulting in plant death. In addition, the petroleum and its derivatives lead to the development of cancer and other diseases. Previous studies indicated that petroleum contamination caused the depression of the nervous system, narcosis and irritation of the mucous membranes of the eyes in humans [8–11]. In view of the high toxicity, carcinogenic, mutagenic and teratogenic potential of petroleum contamination, the bioaccumulation of petroleum in the food chain would disturb biochemical and physiological processes which lead directly or indirectly to human health [12, 13]. Therefore, petroleum contamination is not only a negative impetus for plant growth and development but also an adverse factor for human and ecological health.

2. Technologies for petroleum-contaminated soil remediation

Faced with the serious environmental problems that involve the soil contamination by petroleum, an increasing attention has paid to the development and implementation of innovative technologies for the removal of petroleum from soil in the past decades. Multiple soil remediation technologies involve the physical remediation, chemical remediation and bioremediation were developed and employed for the restoration of petroleum-contaminated soil, particularly the eco-friendly bioremediation (**Figure 1**).

2.1 Physical remediation

Physical remediation uses the physical properties of the contaminants or the contaminated medium to destroy, separate, or contain the contamination, which include soil vapor extraction, flotation, ultrasonication, electro kinetics remediation, thermal desorption and biochar adsorption.

Soil vapor extraction is focus on inducing volatilization of nonaqueous-phase liquid and vapor-phase transport of volatile organic compounds form the subsurface to the surface for subsequent treatment. Soil vapor extraction is also known as in situ soil venting, in situ volatilization, enhanced volatilization, or soil vacuum extraction, in which the extraction well was used to create a pressure or concentration gradient to remove volatiles and some semivolatile contaminants from soil [14]. Soil vapor extraction can remove large quantities of volatile contaminants in uniform soils within a short time. Meanwhile, soil vapor extraction provides oxygen through the flow of air to stimulate the growth of microorganisms. However, the efficiency of soil vapor extraction is affected by soil properties and operational conditions, as well as contaminant properties [15, 16].

The flotation technology relies on the difference in surface properties of both contaminant and soil, which separate oil from soil via a gas–liquid–solid system. The flotation mechanism is dependent on (I) collision between contaminants and

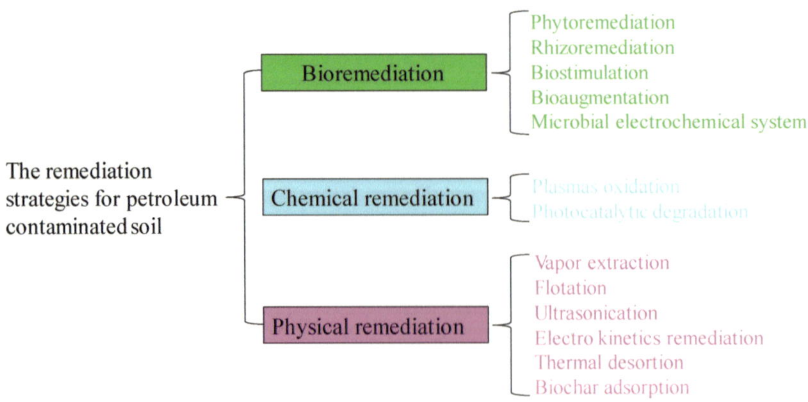

Figure 1.
Schematic of the remediation strategies for petroleum-contaminated soil remediation.

bubbles, (II) form bubble-contaminant with the attachment of contaminants and bubble, (III) flotation of bubble-contaminant based on difference in buoyancy and detachment of contaminant from bubble-contaminant [17]. Flotation is characterized as simplicity, low operational cost and high efficiency for contaminants removal. It also can separate very small or light weight particles with low settling velocities. Nevertheless, large amounts of wastewater were produced during the flotation process. And the efficiency of aged or weathered contaminated soil was significantly decreased [18].

The ultrasonication helps desorption of the contaminant and promotes the formation of strong oxidant, hydroxyl radicals (OH·) which enhance the efficiency of pollutant removal [16, 19]. Ultrasonication can eliminate the hazardous pollutants without any chemicals. Moreover, the on-spot heating and intense agitation will enhances heat and mass transfer processes. However, the higher energy consumption for generation of acoustic makes it the costly setup [20].

Electro kinetics remediation employs direct electric current between appropriately distributed electrodes (cathodes and anodes) that embedded in petroleum-contaminated soil to form an electric field. The voltage potential gradient were formed in the electric field which cause the fluid medium to flow preferentially towards the cathode and drag the contaminant together with the bulk flow [21].The advantage of electro kinetics remediation is speed of execution and low operating cost. Moreover, the electro-osmotic flow is constant through the entire soil mass during the remediation which more suitable for low permeability soils. However, the electro kinetic process is ineffective in low contaminant concentration. The alteration of soil pH and hot spots around the electrodes were induced after an extended period of time [16].

Thermal desorption based on the manipulation of temperatures to increase the vapor pressure of the contaminants, in which the contaminants were volatilized and subsequent desorption from contaminated soil [22]. Thermal desorption is very effective in destroying the oil pollutants under high heat condition. In addition, thermal desorption emits little or no contamination gas into the atmosphere. However, only the volatile contaminants were removed by thermal desorption [23].

Biochar is carbon-enriched and porous with high specific surface area and biodegradability. The biochar was employed as an amendment to implement organic contaminated soil remediation due to the surface adsorption, partition and sequestration [24, 25].

2.2 Chemical remediation

Chemical oxidation has the potential for rapidly deposing or preprocessing soil contaminants. Oxidants that can cause the rapid and complete chemical destruction of petroleum contaminants are employed in the chemical oxidation remediation. In chemical oxidation, the contaminants is oxidation chemically converts to non-hazardous, at least, biodegradable products or less toxic compound that more stable, less mobile or inert [16, 26].

Plasmas oxidation is also considered as highly competitive technology to remediating the pollutants from soils. Especially the plasma technology based on pulsed corona discharge and dielectric barrier discharge has aroused widely concerns in soil remediation [27, 28]. Plasma are macroscopically electrically neutral aggregates composed of numerous ions, electrons, atoms, molecules and unionized neutral particles. A number of active constituents such as O_3, H_2O_2, the hydroxyl radicals (OH·) and high energy electrons were generated in the generation of plasma by ionization, in which a strong oxidizing environment was created for oxidative decomposition of contaminants [29].

Photocatalytic degradation is effective for decomposition of polycyclic aromatic hydrocarbon in the soils. This technology makes use of the semiconductor metal oxide as catalyst to degrading organic pollutant into small molecules directly [30]. Semiconductor molecule contains a valance band with stable energy electrons and an empty higher energy conduction band. The absorption of radiation can initiate the photocatalytic reaction, in which the formation of holes (h^+) in valence band and electrons (e^-) in conduction band in femtosecond time scale. During the photocatalytic process, the hydroxyl radicals (OH·) and superoxide radical anion ($O_2\cdot^-$) are formed to degrading organic pollutant. However, the light absorption characteristics, humic substances content and moisture content of soil may affect the photocatalytic degradation [31, 32].

2.3 Bioremediation

The physical and chemical remediation have their own characteristics, even effective for higher contaminants removal. However, the practical applications are impracticable under some circumstances, such as the remediation of amount of contaminated soil is economically impracticable. Furthermore, most of the physical and chemical remediation technologies are unavoidable to destroy the soil microbiota that reduces the concentration of soil at the expense of damaging the integrity of soil ecosystem [33]. Therefore, alternative technologies that have less environmentally aggressive, greater ease of practical application, as well as more efficient and cost-effective for environmental decontamination are expected. In the past decades, significant advances on bioremediation have been achieved. Although it is time consuming, bioremediation techniques, due to their eco-friendly approach and very low cost, efficient and sustainable for restoring the contaminated soil in the context of sustainability, are extensively noticeable at present [34].

Bioremediation is a process that naturally or artificially take advantage of living organisms or their products to reduce (degrade, detoxify, mineralize or transform) the pollutants of the contaminated environment [35]. For this purpose, living organisms (plants and microorganisms) that tolerate and have capacity to grown under contaminated soil are usually used. Number of studies has revealed that selecting petroleum-tolerant plants for bioremediation in cases of soil petroleum pollution is a feasible and sustainable technology. Many plants, such as perennial ryegrass, alfalfa, *Mirabilis jalapa*, were considered as tolerant to petroleum stress [36–38]. The microorganisms that are utilized in petroleum pollutants removal can be bacteria, fungi or yeasts. These microbes are the essential component in soil ecological systems that play a vital role for the remediation of petroleum hydrocarbon and other pollutants [39, 40]. Some of them have high capacity to degrade contaminants and widely used for environmental depollution [41]. In the bioremediation of petroleum contaminated soils, the most widely used organisms are bacteria which have high frequency, rapid growth and a broad spectrum of degradation of petroleum products [42].

Phytoremediation is a kind of bioremediation. Phytoremediation is considered as an alternative technology that makes use of plants and microorganisms associated with their root to degrade or reduce soil contaminants [43]. The main factors to consider when choosing a plant as a phytoremediator are root system, plant survival and its adaptability to prevailing environmental conditions. Plants not only can degrade petroleum pollutants directly via enzymatic activities, but also can stimulate the rhizosphere microbial community to degrade petroleum contaminants [44]. The use of plant growth promoting rhizobacteria (PGRR) plays an important role in phytoremediation. The PGRR promote the growth of plant by providing phytohormones and mineral nutrition. PGRR can also generate antibiotics, compete for nutrients with pathogens or induce systemic resistance in the host plant to protect it from pathogens [45]. To understand the interactions between PGRR and plants would be better reveal the alleviation of contaminants toxicity of soil.

Rhizoremediation is a strategy for phytoremediation, plant act indirectly in phytoremediation, since their presence in the environment provides favorable conditions for the growth of microorganisms in the rhizosphere region. The rhizosphere is a soil zone ranging from the surface to a depth of 1–5 mm, in which the interdependence between plants and microorganisms result in a symbiotic lead to form a symbiotic relationship [46]. Plants roots can release the organic acids, carbohydrates, amino acid and oxygen to the rhizosphere, which promote the development of rhizosphere microbe (including bacteria, fungi, protists, nematodes and in vertebrates). And the microbe benefits the plant by providing the necessary vitamins, cytokinins and amino acids to promote plant growth [16]. The development of microbial biotechnology is beneficial for screening and identifying microorganisms from petroleum contaminated soils [47]. Many microorganisms have been isolated and utilized as biodegraders for petroleum hydrocarbons disposal. More than 79 genera of bacteria that capable of degrading petroleum hydrocarbons have been identified [48]. Furthermore, some microorganisms were crucial for petroleum hydrocarbons since the abundance of these microorganisms were dominant increased after petroleum contamination [49]. In view of different indigenous bacteria have different catalytic enzymes, the combination of multiple functional bacteria were preferable to remediating the pollutants in contaminated soils. Previous studies showed that the joint action of indigenous bacterial consortium and exogenous bacteria were effectively accelerating the degradation of petroleum [50].

Biostimulation is one of the main strategy bioremediation for the decontamination of petroleum-polluted soil, which through adjusting the environmental conditions (temperature, moisture, pH, redox potential, aeration, mineral nutrition) to enhance the growth and the metabolic activity of indigenous degrading microbial populations. The microorganisms' activity in biostimulation practice is tolerant to various hydrocarbons and can utilized hydrocarbons as carbon sources for their growth [51].

Bioaugmentation is another strategy of bioremediation, which refer to the inoculation of exogenous microorganisms into the contaminated soils to degrade the target contaminants [52]. The inoculated microorganism can be one strain or a consortium of microbial strains with diverse functional degradation capacities [53]. Bioaugmentation was considered to be more effective for the degradation of the light fraction (C12-C23) of petroleum hydrocarbons [54]. Bioaugmentation can divided into cell bioaugmentation and genetic bioaugmentation based on the degradation mechanism of the inoculated strains. Cell bioaugmentation relies on the survival and catabolic activity of the inoculated strains to accelerate the degradation of target contaminants directly [55]. While genetic bioaugmentation based on the spread of catabolic genes (plasmids, integrons or transposons mediated) into native microbial populations. And then the native acquiring these genes achieve the ability to degrade organic contaminants [56]. As compare to the cell bioaugmentation, genetic bioaugmentation, especially plasmid-mediated bioaugmentation appears to have greater potential for the bioremediation of contaminated soils [57].

In addition, microbial electrochemical system was considered as an emerging technique for bioremediation, which integrates microbial and electrochemical processes to convert the pollutants to less-toxic or value-added products [58]. With various inherent advantages, microbial electrochemical system was mostly applied in remediation of petroleum contaminants in soil. Microbial electrochemical system is considered as more flexible for various contaminants in bioremediation due to the oxidation and reduction transformation in remediation processes [59].

3. Influence factors of bioremediation

Bioremediation is viewed as a technique to accelerate the natural biodegradation process in a cost-effective and environment friendly way. However, bioremediation is time consuming, and the contaminant concentration and composition, temperature, soil pH, oxygen condition and salinity are highly affected the bioremediation of the petroleum-contaminated soils.

The plants and microbes would unable to grow in a high petroleum soil. In that case, the bioremediation was inoperative or low efficiency. In addition, some of the petroleum derivatives with high solubility have higher cytotoxicity to biodegradation bacterial, while other compounds produced no significant inhibitory effects on bacterial growth [60].

The temperature plays a vital role in bioremediation and influence biodegradation reactions [61]. Indeed, temperature can indirectly affecting biodegradation efficiency by affecting bacterial growth and metabolism, altering soil matrix and the mode of occurrence of pollutants [62].

Petroleum and its derivatives can full fill the interstices of soil, which reduce the amount of oxygen in soils. Under reduced or absent oxygen conditions, the metabolism of aerobic microorganisms was partially interrupted, as well as the bioavailability and degradation efficiency of pollutant were reduced [12, 63].

The key components of biodegradation of petroleum hydrocarbons are various specific enzymes [61]. The alterations of pH value may influence the enzymes activities to reduce the effective of the biodegradation. In other hand, alterations of pH and high salinity may inhibit microbial growth and metabolism. In addition, the lack of technique to monitor the survival and activity of the organism in soil also limited the application of bioremediation.

4. Conclusions

Bioremediation is an eco-friendly and economic method to remove the petroleum pollutants of soil. There were several kinds of bioremediation have been applied to remediate the petroleum hydrocarbon from contaminated soils, such as phytoremediation, rhizoremediation, biostimulation, bioaugmentation, and so on. The petroleum-tolerant plants and the high-effective petroleum degradation microbes are preferable used in bioremediation process. Alternatively, construction of the different functional bacterial consortium or genetic engineering bacteria, and potentially using integrated bioremediation approaches for bioremediation of petroleum have become a trend in future. However, some of the influence factors are significant reduced the degradation efficiency on bioremediation application. Therefore, use a combination of bioremediation and other technologies is effective strategy to accelerate for petroleum hydrocarbon pollutants removal. Furthermore, developing and enriching the novel technologies of bioremediation remains far off.

Acknowledgements

This research was supported by Scientific Research Foundation for the Introduction of Talent of Shenyang Agricultural University (Program No. 88481607).

Conflict of interest

The author declares no conflict of interest.

Author details

Shuisen Chen* and Ming Zhong
Key Laboratory of Agricultural Biotechnology of Liaoning Province, College of Bioscience and Biotechnology, Shenyang Agricultural University, Shenyang, China

*Address all correspondence to: shuisenchen@syau.edu.cn

References

[1] Sun Y. On-site management of International Petroleum Cooperation Projects. Natural Gas Exploration and Development. 2009;**32**(2):69-73. DOI: 10.3969/j.issn.1673-3177.2009.02.019

[2] Falkova M, Vakh C, Shishov A, Zubakina E, Moskvin A, Moskvin L, et al. Automated IR determination of petroleum products in water based on sequential injection analysis. Talanta. 2016;**148**:661-665. DOI: 10.1016/j.talanta.2015.05.043

[3] Smith E, Thavamani P, Ramadass K, Naidu R, Srivastava P, Megharaj M. Remediation trials for hydrocarbon-contaminated soils in arid environments: Evaluation of bioslurry and biopiling techniques. International Biodeterioration & Biodegradation. 2015;**101**:56-65. DOI: 10.1016/j.ibiod.2015.03.029

[4] Redondo-Gómez S, Petenello MC, Feldman SR. Growth, nutrient status, and photosynthetic response to diesel-contaminated soil of a cordgrass, *Spartina argentinensis*. Marine Pollution Bulletin. 2014;**79**:34-38. DOI: 10.1016/j.marpolbul.2014.01.009

[5] Rahbar FG, Kiarostami K, Shirdam R. Effects of petroleum hydrocarbons on growth, photosynthetic pigments and carbohydrate levels of sunflower. Journal of Food, Agriculture and Environment. 2012;**10**:773-776

[6] Nardeli SM, Saad CF, Rossetto P de B, Caetano VS, Ribeiro-Alves M, Paes JES, et al. Transcriptional responses of *Arabidopsis thaliana* to oil contamination. Environmental and Experimental Botany. 2016;**127**:63-72. DOI: 10.1016/j.envexpbot.2016.03.007

[7] Zhang J, Fan S, Yang J, Du X, Li F, Hou H. Petroleum contamination of soil and water, and their effects on vegetables by statistically analyzing entire data set. Science of the Total Environment. 2014;**476-477**:258-265. DOI: 10.1016/j.scitotenv.2014.01.023

[8] Bezza FA, Nkhalambayausi Chirwa EM. Biosurfactant from *Paenibacillus dendritiformis* and its application in assisting polycyclic aromatic hydrocarbon (PAH) and motor oil sludge removal from contaminated soil and sand media. Process Safety and Environmental Protection. 2015;**98**:354-364. DOI: 10.1016/j.psep.2015.09.004

[9] Ameen F, Moslem M, Hadi S, Al-Sabri AE. Biodegradation of diesel fuel hydrocarbons by mangrove fungi from Red Sea Coast of Saudi Arabia. Saudi Journal of Biological Sciences. 2016;**23**:211-218. DOI: 10.1016/j.sjbs.2015.04.005

[10] Bastida F, Jehmlich N, Lima K, Morris BEL, Richnow HH, Hernández T, et al. The ecological and physiological responses of the microbial community from a semiarid soil to hydrocarbon contamination and its bioremediation using compost amendment. Journal of Proteomics. 2016;**135**:162-169. DOI: 10.1016/j.jprot.2015.07.023

[11] Al Shami A, Harik G, Alameddine I, Bruschi D, Garcia DA, El-Fadel M. Risk assessment of oil spills along the Mediterranean coast: A sensitivity analysis of the choice of hazard quantification. Science of the Total Environment. 2017;**574**:234-245. DOI: 10.1016/j.scitotenv.2016.09.064

[12] Chen M, Xu P, Zeng G, Yang C, Huang D, Zhang J. Bioremediation of soils contaminated with polycyclic aromatic hydrocarbons, petroleum, pesticides, chlorophenols and heavy metals by composting: Applications, microbes and future research needs. Biotechnology Advances.

2015;**33**:745-755. DOI: 10.1016/j.biotechadv.2015.05.003

[13] Rein A, Adam IKU, Miltner A, Brumme K, Kästner M, Trapp S. Impact of bacterial activity on turnover of insoluble hydrophobic substrates (phenanthrene and pyrene)-model simulations for prediction of bioremediation success. Journal of Hazardous Materials. 2016;**306**:105-114. DOI: 10.1016/j.jhazmat.2015.12.005

[14] Baker RS, Bierschenk J. Vacuum-enhanced recovery of water and NAPL: Concept and field test. Soil and Sediment Contamination. 1995;**4**:57-76

[15] Albergaria JT, Alvim-Ferraz M d CM, Delerue-Matos C. Remediation of sandy soils contaminated with hydrocarbons and halogenated hydrocarbons by soil vapour extraction. Journal of Environmental Management. 2012;**104**:195-201. DOI: 10.1016/j.jenvman.2012.03.033

[16] Lim MW, Lau EV, Poh PE. A comprehensive guide of remediation technologies for oil contaminated soil—Present works and future directions. Marine Pollution Bulletin. 2016;**109**:14-45. DOI: 10.1016/j.marpolbul.2016.04.023

[17] Tao D. Role of bubble size in flotation of coarse and fine particles—A review. Separation Science and Technology. 2005;**39**:741-760. DOI: 10.1081/SS-120028444

[18] Wang J, Yin J, Ge L, Zheng J. Using flotation to separate oil spill contaminated beach sands. Journal of Environmental Engineering. 2010;**136**:147-151. DOI: 10.1061/(ASCE)EE.1943-7870.0000117

[19] Zhang J, Li J, Thiring RW, Hu X, Song X. Oil recovery from refinery oily sludge via ultrasound and freeze/thaw. Journal of Hazardous Materials. 2012;**203-204**:195-203. DOI: 10.1016/j.jhazmat.2011.12.016

[20] Son Y, Nam S, Ashokkumar M, Khim J. Comparison of energy consumptions between ultrasonic, mechanical, and combined soil washing processes. Ultrasonics Sonochemistry. 2012;**19**:395-398. DOI: 10.1016/j.ultsonch.2011.11.002

[21] Ranjan RS, Qian Y, Krishnapillai M. Effects of electrokinetics and cationic surfactant cetyltrimethylammonium bromide [CTAB] on the hydrocarbon removal and retention from contaminated soils. Environmental Technology. 2006;**27**:767-776. DOI: 10.1080/09593332708618686

[22] Rushton DG, Ghaly AE, Martinell K. Assessment of Canadian regulations and remediation methods for diesel oil contaminated soils. American Journal of Applied Sciences. 2007;**4**:465-578. DOI: 10.3844/ajassp.2007.465.478

[23] Falciglia PP, Giustra MG, Vagliasindi FGA. Low-temperature thermal desorption of diesel polluted soil: Influence of temperature and soil texture on contaminant removal kinetics. Journal of Hazardous Materials. 2011;**185**:392-400. DOI: 10.1016/j.jhazmat.2010.09.046

[24] Tang J, Zhu W, Kookana R, Katayama A. Characteristics of biochar and its application in remediation of contaminated soil. Journal of Bioscience and Bioengineering. 2013;**116**:653-659. DOI: 10.1016/j.jbiosc.2013.05.035

[25] Wu S, He H, Inthapanya X, Yang C, Lu L, Zeng G, et al. Role of biochar on composting of organic wastes and remediation of contaminated soils—A review. Environmental Science and Pollution Research International. 2017;**24**:16560-16577. DOI: 10.1007/s11356-017-9168-1

[26] Jalilian Ahmadkalaei SP, Gan S, Ng HK, Abdul Talib S. Evaluation of ethyl lactate as solvent in Fenton

oxidation for the remediation of total petroleum hydrocarbon (TPH)-contaminated soil. Environmental Science and Pollution Research International. 2017;**24**:17779-17789. DOI: 10.1007/s11356-017-9382-x

[27] Wang TC, Qu G, Li J, Liang D. Evaluation of the potential of soil remediation by direct multi-channel pulsed corona discharge in soil. Journal of Hazardous Materials. 2014;**264**:169-175. DOI: 10.1016/j.jhazmat.2013.11.011

[28] Ognier S, Rojo J, Liu Y, Duten X, Cavadias S, Thannberger L. Mechanisms of pyrene degradation during soil treatment in a dielectric barrier discharge reactor. Plasma Processes and Polymers. 2014;**11**:734-744. DOI: 10.1002/ppap.201300077

[29] Li R, Liu Y, Mu R, Cheng W, Ognier S. Evaluation of pulsed corona discharge plasma for the treatment of petroleum-contaminated soil. Environmental Science and Pollution Research International. 2017;**24**:1450-1458. DOI: 10.1007/s11356-016-7929-x

[30] Dong D, Li P, Li X, Zhao Q, Zhang Y, Jia C, et al. Investigation on the photocatalytic degradation of pyrene on soil surfaces using nanometer anatase TiO_2 under UV irradiation. Journal of Hazardous Materials. 2010;**174**:859-863. DOI: 10.1016/j.jhazmat.2009.09.132

[31] Hoffmann MR, Martin ST, Wonyong C, Bahnemann DW. Environmental applications of semiconductor photocatalysis. Chemical Reviews. 1995;**95**:69-96. DOI: 10.1021/cr00033a004

[32] Cheng M, Zeng G, Huang D, Lai C, Xu P, Zhang C, et al. Hydroxyl radicals based advanced oxidation processes (AOPs) for remediation of soils contaminated with organic compounds: A review. Chemical Engineering Journal. 2016;**284**:582-598. DOI: 10.1016/j.cej.2015.09.001

[33] Gómez-Sagasti MT, Epelde L, Alkorta I, Garbisu C. Reflections on soil contamination research from a biologists point of view. Applied Soil Ecology. 2016;**105**:207-210. DOI: 10.1016/j.apsoil.2016.04.004

[34] Azubuike CC, Chikere CB, Okpokwasili GC. Bioremediation techniques-classification based on site of application: Principles, advantages, limitations and prospects. World Journal of Microbiology and Biotechnology. 2016;**32**:180. DOI: 10.1007/s11274-016-2137-x

[35] Gouda S, Kerry RG, Das G, Paramithiotis S, Shin H-S, Patra JK. Revitalization of plant growth promoting rhizobacteria for sustainable development in agriculture. Microbiological Research. 2018;**206**:131-140. DOI: 10.1016/j.micres.2017.08.016

[36] Günther T, Dornberger U, Fritsche W. Effects of ryegrass on biodegradation of hydrocarbons in soil. Chemosphere. 1996;**33**:203-215

[37] Peng S, Zhou Q, Cai Z, Zhang Z. Phytoremediation of petroleum contaminated soils by *Mirabilis jalapa* L. in a greenhouse plot experiment. Journal of Hazardous Materials. 2009;**168**:1490-1496. DOI: 10.1016/j.jhazmat.2009.03.036

[38] Shahsavari E, Adetutu EM, Anderson PA, Ball AS. Tolerance of selected plant species to petrogenic hydrocarbons and effect of plant rhizosphere on the microbial removal of hydrocarbons in contaminated soil. Water, Air, and Soil Pollution. 2013;**224**:1495. DOI: 10.1007/s11270-013-1495-3

[39] Yan L, Sinkko H, Penttinen P, Lindström K. Characterization of successional changes in bacterial community composition during bioremediation of used motor oil-contaminated soil in a boreal climate.

Science of the Total Environment. 2016;**542**:817-825. DOI: 10.1016/j.scitotenv.2015.10.144

[40] Jia J, Zong S, Hu L, Shi S, Zhai X, Wang B, et al. The dynamic change of microbial communities in crude oil-contaminated soils from oil fields in China. Soil and Sediment Contamination: An International Journal. 2017;**26**:171-183. DOI: 10.1080/15320383.2017.1264923

[41] Kuppusamy S, Thavamani P, Venkateswarlu K, Lee YB, Naidu R, Megharaj M. Remediation approaches for polycyclic aromatic hydrocarbons (PAHs) contaminated soils: Technological constraints, emerging trends and future directions. Chemosphere. 2017;**168**:944-968. DOI: 10.1016/j.chemosphere.2016.10.115

[42] Wolicka D, Suszek A, Borkowski A, Bielecka A. Application of aerobic microorganisms in bioremediation in situ of soil contaminated by petroleum products. Bioresource Technology. 2009;**100**:3221-3227. DOI: 10.1016/j.biortech.2009.02.020

[43] Brzeszcz J, Steliga T, Kapusta P, Turkiewicz A, Kaszycki P. R-strategist versus K-strategist for the application in bioremediation of hydrocarbon-contaminated soils. International Biodeterioration & Biodegradation. 2016;**106**:41-52. DOI: 10.1016/j.ibiod.2015.10.001

[44] Hall J, Soole K, Bentham R. Hydrocarbon phytoremediation in the family Fabaceae—A review. International Journal of Phytoremediation. 2011;**13**:317-332. DOI: 10.1080/15226514.2010.495143

[45] Jéssica Janzen dos S, Maranho LT. Rhizospheric microorganisms as a solution for the recovery of soils contaminated by petroleum: A review. Journal of Environmental Management. 2018;**210**:104-113. DOI: 10.1016/j.jenvman.2018.01.015

[46] Newman MM, Lorenz N, Hoilett N, Lee NR, Dick RP, Liles MR, et al. Changes in rhizosphere bacterial gene expression following glyphosate treatment. Science of the Total Environment. 2016;**553**:32-41. DOI: 10.1016/j.scitotenv.2016.02.078

[47] Guerra AB, Oliveira JS, Silva-Portela RCB, Araújo W, Carlos AC, Vasconcelos ATR, et al. Metagenome enrichment approach used for selection of oil-degrading bacteria consortia for drill cutting residue bioremediation. Environmental Pollution. 2018;**235**:869-880. DOI: 10.1016/j.envpol.2018.01.014

[48] Tremblay J, Yergeau E, Fortin N, Cobanli S, Elias M, King TL, et al. Chemical dispersants enhance the activity of oil- and gas condensate-degrading marine bacteria. The ISME Journal. 2017;**11**:2793-2808. DOI: 10.1038/ismej.2017.129

[49] Yakimov MM, Timmis KN, Golyshin PN. Obligate oil-degrading marine bacteria. Current Opinion in Biotechnology. 2007;**18**:257-266. DOI: 10.1016/j.copbio.2007.04.006

[50] Tao K, Liu X, Chen X, Hu X, Cao L, Yuan X. Biodegradation of crude oil by a defined co-culture of indigenous bacterial consortium and exogenous *Bacillus subtilis*. Bioresource Technology. 2017;**224**:327-332. DOI: 10.1016/j.biortech.2016.10.073

[51] Shahi A, Aydin S, Ince B, Ince O. Evaluation of microbial population and functional genes during the bioremediation of petroleum-contaminated soil as an effective monitoring approach. Ecotoxicology and Environmental Safety. 2016;**125**:153-160. DOI: 10.1016/j.ecoenv.2015.11.029

[52] Ruffini Castiglione M, Giorgetti L, Becarelli S, Siracusa G, Lorenzi R, Di Gregorio S. Polycyclic aromatic

hydrocarbon-contaminated soils: Bioaugmentation of autochthonous bacteria and toxicological assessment of the bioremediation process by means of *Vicia faba* L. Environmental Science and Pollution Research International. 2016;**23**:7930-7941. DOI: 10.1007/s11356-016-6049-y

[53] Heinaru E, Merimaa M, Viggor S, Lehiste M, Leito I, Truu J, et al. Biodegradation efficiency of functionally important populations selected for bioaugmentation in phenol- and oil-polluted area. FEMS Microbiology Ecology. 2005;**51**:363-373. DOI: 10.1016/j.femsec.2004.09.009

[54] Bento FM, Camargo FAO, Okeke BC, Frankenberger WT. Comparative bioremediation of soils contaminated with diesel oil by natural attenuation, biostimulation and bioaugmentation. Bioresource Technology. 2005;**96**:1049-1055. DOI: 10.1016/j.biortech.2004.09.008

[55] Singh A, Ward OP. Biotechnology and Bioremediation—An overview. In: Singh A, Ward OP, editors. Biodegradation and Bioremediation. Berlin, Heidelberg: Springer; 2004. pp. 1-17. DOI: 10.1007/978-3-662-06066-7_1

[56] Wiedenbeck J, Cohan FM. Origins of bacterial diversity through horizontal genetic transfer and adaptation to new ecological niches. FEMS Microbiology Reviews. 2011;**35**:957-976. DOI: 10.1111/j.1574-6976.2011.00292.x

[57] Garbisu C, Garaiyurrebaso O, EpeldeL,GrohmannE,AlkortaI.Plasmid-mediated bioaugmentation for the bioremediation of contaminated soils. Frontiers in Microbiology. 2017;**8**. DOI: 10.3389/fmicb.2017.01966

[58] Wang X, Cai Z, Zhou Q, Zhang Z, Chen C. Bioelectrochemical stimulation of petroleum hydrocarbon degradation in saline soil using U-tube microbial fuel cells. Biotechnology and Bioengineering. 2012;**109**:426-433. DOI: 10.1002/bit.23351

[59] Wu Y, Jing X, Gao C, Huang Q, Cai P. Recent advances in microbial electrochemical system for soil bioremediation. Chemosphere. 2018;**211**:156-163. DOI: 10.1016/j.chemosphere.2018.07.089

[60] Hou N, Zhang N, Jia T, Sun Y, Dai Y, Wang Q, et al. Biodegradation of phenanthrene by biodemulsifier-producing strain *Achromobacter* sp. LH-1 and the study on its metabolisms and fermentation kinetics. Ecotoxicology and Environmental Safety. 2018;**163**:205-214. DOI: 10.1016/j.ecoenv.2018.07.064

[61] Varjani SJ, Upasani VN. A new look on factors affecting microbial degradation of petroleum hydrocarbon pollutants. International Biodeterioration & Biodegradation. 2017;**120**:71-83. DOI: 10.1016/j.ibiod.2017.02.006

[62] Abed RMM, Al-Kharusi S, Al-Hinai M. Effect of biostimulation, temperature and salinity on respiration activities and bacterial community composition in an oil polluted desert soil. International Biodeterioration & Biodegradation. 2015;**98**:43-52. DOI: 10.1016/j.ibiod.2014.11.018

[63] Martínez Álvarez LM, Lo Balbo A, Mac Cormack WP, Ruberto LAM. Bioremediation of a petroleum hydrocarbon-contaminated Antarctic soil: Optimization of a biostimulation strategy using response-surface methodology (RSM). Cold Regions Science and Technology. 2015;**119**:61-67. DOI: 10.1016/j.coldregions.2015.07.005

Chapter 10

Mycoremediation in Soil

Francesca Bosco and Chiara Mollea

Abstract

The chapter reviews the most important researches on the use of micro- and macrofungi in the bioremediation of contaminated soils. In particular, the main classes of soil pollutants in Europe (heavy metals, mineral oils, polycyclic aromatic hydrocarbons (PAHs), monoaromatic hydrocarbons, phenols and chlorinated hydrocarbons (CHCs)), together with the emerging contaminants (i.e. endocrine-disrupting chemicals (EDCs) and pharmaceutical-personal care products (PPCPs)) are considered. A description of the fungal species (saprotrophic and biotrophic basidiomycetes) and biodegradative extracellular (laccases and class II peroxidases) and intracellular (cytochrome P450 monooxygenases and glutathione transferases) enzyme classes is reported. Moreover, the chemical-physical parameters that influence the biodegradation process are examined, and the biostimulation and bioaugmentation strategies are described. A specific attention is paid to the micro-cosm studies, at the laboratory scale, which are an essential approach to evaluate the feasibility of a biodegradation process.

Keywords: mycoremediation, filamentous fungi, mushroom, microcosm, biostimulation, bioaugmentation, laccases, peroxidases, cytochrome P450 monooxygenases, glutathione transferases

1. Introduction

The contamination of soil, water and air by toxic chemicals represents one of the major worldwide environmental problems. From this point of view, the European Union (EU) is paying attention to the improvement of soil protection and recovery and to the prevention of soil contamination, since there are still many historical and new contaminated sites that require remediation [1, 2]. The main classes of soil pollutants in Europe have been reported in [3].

Bioremediation is a simple and cost-effective method that, in the last decades, has received worldwide a particular attention. The general term "bioremediation" indicates the use of living organisms (i.e. bacteria, fungi, algae and plants) in the detoxification of polluted soils and wastewaters. In a bioremediation process, organic and inorganic hazardous substances may degrade, accumulate or immobilize, resulting in a significant reduction of the contamination level.

In the last decay, the role of fungi in bioremediation has been increasingly recognized [4, 5]. About this, various authors have highlighted the ability of fungi, mainly saprotrophic and biotrophic basidiomycetes, to degrade or to transform toxic compounds [6, 7]. Mycoremediation is the bioremediation technique which employ fungi in the removal of toxic compounds; it could be carried out in the presence of both filamentous fungi (moulds) [8] and macrofungi (mushrooms) [9, 10]. Both classes possess enzymes for the degradation of a large variety of pollutants [11, 12].

Fungi are well known for their ability to colonize a wide range of heterogeneous environments and for their ability to adapt to the complex soil matrices, also at extreme environmental conditions. Furthermore, they can decompose the organic matter and easily colonize both biotic and abiotic surfaces [13, 14].

Filamentous fungi show some peculiar characteristics that make them more advisable in soil bioremediation than yeasts and bacteria [14, 15]. The most important are the type of growth (i.e. the development of a multicellular mycelial network) suited to soil colonization and translocation of nutrients and water, the production of many bioactive compounds and extracellular enzymes and the unique capability to co-metabolize many environmental chemicals [16].

Mycoremediation represents thus a biological tool to degrade, transform or immobilize environmental contaminants.

The state of the art of soil mycoremediation is reviewed in the present chapter. A particular attention is given to the fungal species and enzymes involved in the biodegradation processes, together with the classes of toxic compounds that could be biodegraded. Bioremediation strategies (i.e. biostimulation and bioaugmentation) and significant examples of microcosm and field studies are also discussed. Finally, the application of mushrooms as emerging technology in soil mycoremediation is reported.

2. Important fungal species involved in biodegradation

The most suitable fungi to be used in soil remediation are basidiomycetes and, in particular, the ecological groups of saprotrophic and biotrophic fungi [17].

The saprotrophic basidiomycetes, which use dead organic matter as a carbon source, include the wood-degrading fungi. Among them, white-rot fungi (WRF) are considered for the leading role in biodegradation [18]. WRF can degrade efficiently both lignin and cellulose biopolymers till the complete mineralization [19], thanks to the production of an extracellular enzymatic complex, which comprehend lignin peroxidases (LiPs), manganese-dependent peroxidases (MnPs), versatile peroxidases (VPs), laccases, H_2O_2-generating oxidases and dehydrogenases, produced during the idiophase, usually under nitrogen depletion.

Some of the most representative WRF, able to degrade pollutants, include *Phanerochaete chrysosporium*, *Pleurotus ostreatus*, *Trametes versicolor*, *Bjerkandera adusta*, *Lentinula edodes*, *Irpex lacteus*, *Agaricus bisporus*, *Pleurotus tuber-regium* and *Pleurotus pulmonarius* [20, 21]. Among these fungi, *Phanerochaete chrysosporium* has been the most investigated for its ability to degrade toxic or insoluble compounds to CO_2 and H_2O, more efficiently than other fungi. In 1985, for the first time, Bumpus et al. proposed the application of *Phanerochaete chrysosporium* in bioremediation studies, and the fungus became a model system in the mycoremediation field [22].

The biotrophic basidiomycetes comprehend ectomycorrhizas which obtain the carbon source from a mutualistic plant partner: the fungal hyphal network envelopes the root and penetrates between the cells of the root cortex [17]. Ectomycorrhizal fungi (ECM) can assemble and recycle the nutrients from the organic matter of the soil [23]. ECM comprehends about 10,000 fungal species; the most representatives are *Amanita* spp., *Boletus* spp., *Gautieria* spp., *Hebeloma* spp., *Lactarius* spp., *Morchella* spp., *Suillus* spp. and *Rhizopogon* spp. [16, 24]. ECM fungi secrete enzymes to get nutrients by means of the degradation of molecules in the soil organic matter. ECM possesses extracellular and cytosolic enzymes which attack molecules containing N and P atoms. Hydrolytic enzymes comprehend β-glucosidases and phosphatases, while the oxidative ones are peroxidases and phenol oxidases [25]. The ECM fungi application is important in habitat where the litter layer is restricted and consequently ligninolytic enzymes, typical of wood fungi, are not so efficacious; in this contest, ECM fungi are able to produce enzymes to sequester nutrients directly from the soil. The same enzymes allow ECM to degrade many persistent organic pollutants [26].

Most of the biodegradation studies at the laboratory and field scale are concerned to microfungi, but in the last years, much attention has been given to mushrooms which are broadly present in soil and also easily soil-cultivated [27]. Bioremediation by macrofungi basidiomycetes is reported by [28] to be advantageous because, together with remediation, soil is enriched with organic matter and nutrients and plant growth results enhanced. These macrofungi are potent degraders thanks to the secretion of the same non-specific enzymes (LiP, MnP and laccase) described for the wood-degrading fungi and, for this reason, are interesting in the bioremediation field. At the same time, they grow to a great extent producing high biomass quantities, in particular when cultivated on carbon sources, such as straw or sawdust [29]. The mushroom biomass can be a protein source or can contain biologically active compounds such as phenols with antioxidant activity [12, 30]. Furthermore, mushroom biomass can be applied in biosorption treatment thanks to its ability to accumulate ions and xenobiotics from contaminated soils [31].

3. Toxic compounds degraded by fungi

The biodegradation capability of different hydrocarbon classes such as mineral oils, polycyclic aromatic hydrocarbons (PAHs), monoaromatic hydrocarbons and chlorinated hydrocarbons (CHCs), together with phenols, was demonstrated for many fungal species [17]. Moreover, the possibility to decrease the risk associated with heavy metals, metalloids and radionuclides in soil has been described [16].

Cd, Cr, Hg, Pb, Cu, Zn and As are the most common heavy metals found in soil. In the EU, more than 80,000 contaminated sites are counted. Heavy metals can be generated by natural processes, like the metal-enriched rock erosion, and anthropogenic activities (e.g. mining, smelting, fossil fuel combustion, waste disposal, corrosion and agricultural practices) [32, 33]. Heavy metals that enter the environment can be transported or transformed by means of photo-, chemical- or biodegradation; moreover, they can also be biotransformed [34]. Fungi are potential heavy metal accumulators; in particular basidiomycetes mushrooms can uptake heavy metals from soil by means of their mycelia and accumulate them in the fruiting bodies, irrespective of their age [35]. As reported by [10], species of *Agaricus*, *Amanita*, *Cortinarius*, *Boletus*, *Leccinum*, *Suillus* and *Phellinus* are some of the mushroom applicable for the mobilization/complexation of different heavy metals in soil.

In the EU, mineral oils, together with heavy metals, represent the main source of soil contamination, significantly greater than 60% of the total contaminants. Mineral oils, refined from crude petroleum oil, are a group of various hydrocarbons, straight and branched-chain paraffinic, naphthenic and aromatic ones, with 15 or more C numbers [2]. They can be used for the preparation of lubricant products (e.g. engine oils or hydraulic fluids) or "non-lubricant" ones (e.g. agricultural spray oils). Their industrial application is at a large scale, and the soil contamination can occur during transport, storage or refining or also for accidental leakages [36]. Hydrolases, dehydrogenases and membrane-bound cytochrome P450 enzymes constitute the fungal hydrocarbon-degrading system [37]. Fungal species belonging to *Rhizopus*, *Paecilomyces*, *Alternaria*, *Mucor*, *Gliocladium*, *Aspergillus*, *Fusarium*, *Cladosporium*, *Geotrichum*, *Penicillium* and *Pleurotus* are capable of utilizing crude oil as the sole carbon and energy source [37–39].

Polycyclic aromatic hydrocarbons (PAHs), molecules with multiple carbon rings, derive from the incomplete combustion of organic materials. Their origin can be both natural (e.g. open burning, natural losses of petroleum and volcanic activities) and predominantly anthropogenic (e.g. residential heating, coal gasification, carbon black, activities in petroleum refineries). PAH contamination corresponds to 13%: these compounds tend to bound to soil particles and to remain absorbed [40]. Both ligninolytic and non-ligninolytic fungi are able to degrade PAHs by means of the extracellular lignin-degrading enzymatic system, which contribute to the first attack on PAHs, and of the P450 monooxygenase [41]. Apart from the model *P. chrysosporium*, species belonging to *Aspergillus*, *Penicillium*, *Rhizopus*, *Fusarium*, *Cladosporium* and *Trichoderma* are capable of degrading PAHs [42].

Another group of crude petrol-derived hydrocarbons, which represent the 6% of soil contaminants, is that of monoaromatic hydrocarbons, and in particular those grouped in the acronym BTEX (benzene, toluene, ethylbenzene and xylene). Fungi are efficient in aromatic hydrocarbon degradation, as for PAH degradation, thanks to the ligninolytic enzymatic system. WRF, such as *P. chrysosporium* and *Trametes versicolor*, are reported to be good BTEX degraders together with soil and mycorrhizal fungi [43].

Phenols consist of one or more aromatic rings with hydroxyl functional groups; they are present in the waste streams of almost all the phenolic-using industries (e.g. chemical, paper, food and textile industries) and contaminate the soil as leachates or particulate matter [44, 45]. The percentage of soil contamination is one of the lowest, being around 4% [33]. The biodegradation of phenols is mainly concerned to the production of phenol oxidase enzymes (laccases, tyrosinases and peroxidases) by basidiomycetes: they act on phenols and incorporate one or two atoms of oxygen [46, 47]. Due to the production of these multiple oxidative enzymes, *Trametes* spp., *Lentinus* spp., *Pleurotus* spp. and *Ganoderma* spp. are some of the most efficacious fungal species involved in phenol compound biodegradation [48].

The soil contamination of CHCs is about 2%. These compounds contain Cl atoms substituted for hydrogen atoms normally bonded to a carbon. This group of chemicals comprehends highly toxic pollutants such as polychlorinated biphenyls (PCBs) and chlorinated pesticides, e.g., DDT [49]. As for PAH biodegradation, WRF have been intensively proposed as biodegraders of CHCs due to their unspecific oxidative enzymes. However, also non-WRF, in particular soil ascomycetes and zygomycetes, are able to enzymatically transform these pollutants; in particular, they have the advantage over WRF to tolerate neutral pH and adverse growth conditions [50].

In the last years, emerging contaminants have become of great interest [51]. Among them, the anthropogenic chemicals, endocrine-disrupting chemicals (EDCs) and pharmaceutical-personal care products (PPCPs) are relevant due to their biological effects on nontarget organisms; in particular, EDCs simulate or antagonize the endogenous hormone effects and are toxic to organisms also at very low concentrations. Estrone, 17β-estradiol, 17α-ethinylestradiol, bisphenol A and triclosan are the most detected and studied in soil. EDCs and PPCPs mainly enter the soil environment via irrigation with contaminated wastewater [52–54]. As reviewed by [55], ligninolytic fungi are able to transform EDCs allowing a reduction of the endocrine-disrupting activity or their ecotoxicity; moreover, these fungi are also reported to be able to degrade the heterogeneous class of PPCPs thanks to their broadly unspecific enzymatic systems [56].

4. Enzymes involved in biodegradation of toxic compounds

Since 1985, after the discovery of Bumpus [22] about the degradation potentialities of *P. chrysosporium*, a plethora of authors have described the fungal enzymatic machinery and its role in the transformation of a wide range of organic pollutants in soils. Most of the enzymes are extracellular and allow to attack and then degrade large molecules into smaller units which can enter the cells for further transformations [57].

Extracellular laccases start ring cleavage in the biodegradation of aromatic compounds [8]. They are multicopper oxidases with low substrate specificity and can act on o- and p-phenols, aminophenols and phenylenediamines thanks to a four-electron transfer from the organic substrate to molecular oxygen. The laccase-mediator systems (LMSs) have an effect on the electron transfer chain increasing the laccase substrate range [58].

Fungal peroxidases generate oxidants which initiate the substrate oxidation in the extracellular environment [8]. They belong to the class II peroxidases [59] and catalyse the oxidative conversion of various compounds utilizing H_2O_2 as electron acceptor. As previously reported, LiPs, MnPs and VPs are the main fungal high-redox class II peroxidases. They are involved in the biodegradation of the complex lignocellulose structure and, consequently, can degrade various organic substrates and transform some inorganic ones [46]. Fungi can also secrete the dye-decolorizing peroxidases (DyPs), which have oxidative and hydrolytic activities on phenolic and non-phenolic organic compounds [60]. Heme-thiolate peroxidases (HTPs) transfer peroxide-oxygen, from H_2O_2 or R-COOH to substrate molecules; in this group chloroperoxidases (CPOs) and the unspecific or aromatic peroxygenases (UPOs or APOs) are included. In particular, UPOs can mainly operate on heterogeneous substrates thanks to aromatic peroxygenation, double-bond epoxidation or hydroxylation of aliphatic compounds [59].

Intracellular detoxification pathways comprehend multigenic families of cytochrome P450 monooxygenases and glutathione transferases, mainly owned by wood and plant litter fungi but also by some symbiotic species [46]. These intracellular enzymes have functional roles in fungal primary and secondary metabolism.

P450 cytochrome monooxidases, heme-thiolate-containing oxidoreductases, can act on various substrates in stereo- and regioselective manner, needing O_2 for the reaction. They are activated by a reduced heme iron and add one atom of molecular oxygen to a substrate. Hydroxylation, epoxidation, sulfoxidation and dealkylation can occur and require NAD(P)H as electron donor [61].

Glutathione transferases are located in different cellular compartments and catalyse the nucleophilic attack of an electrophilic C, N or S atom in non-polar compounds by means of reduced glutathione (GSH). When electrophilic substrates are conjugated with GSH, they become more water-soluble. These enzymes have a wide substrate specificity and take part in the detoxification of different endogenous toxic metabolites and exogenous toxic chemicals [62].

5. Main parameters that influence mycoremediation

In general, chemical-physical characteristics of soil, such as pH, temperature, water content and redox potential, show a significant impact on the microbial growth and consequently on the success of a bioremediation process.

In particular, the biodegradation activity of the microorganisms depends on macro- and micronutrient availability in soil and on the presence of any other factor that influence the microbial metabolism, such as the contaminant type and concentration, and their bioavailability, toxicity and mobility [33].

A proper amount of nutrients for microbial growth is usually present in soil; nevertheless, nutrients can also be added in a functional form which serves as an electron donor to stimulate bioremediation process [63]. The biodegradation of a toxic compound mainly depends on the genetic characteristics of the microorganism, in particular on both the extracellular and intracellular enzymatic systems [64]. The contaminant concentration directly influences the microbial activity: a high concentration may produce a variety of toxic effects on the different microbial classes, whereas a low concentration could not be enough to activate degradative enzyme synthesis. Filamentous fungi, able to form extended mycelial network and to synthetize a lot of aspecific enzymes, generally show a higher resistance to high contaminant concentration than bacteria [16]. Moreover, thanks to the low substrate specificity, the synthesis of degradative enzymes occurs also at low contaminant concentrations. The intracellular metabolic pathways involved in mycoremediation show remarkable similarities with those that regulate the secondary metabolism in fungi, in particular those of mycotoxin production [64]. Filamentous fungi which produce mycotoxins (e.g. *Aspergillus* and *Penicillium* spp.) exhibit the ability to degrade a wide variety of pharmaceutical compounds [65], among them the emerging pollutants EDCs [16, 66], ineffectively degraded by bacteria. The contaminant bioavailability is one of the most important factors that can be improved to optimize and accelerate the biodegradation; this fact has been demonstrated in the mycoremediation of aged PAH-contaminated soils [67]. The fungal ability to chemically modify or affect the contaminant bioavailability by means of biosurfactant production has been reported in different reviews [68, 69]. *Penicillium* and *Aspergillus* species have been reported to be biosurfactant producers [70, 71]. A wide range of microbial biosurfactant applications have been reported in the environmental protection field (e.g. enhancing oil recovery, controlling oil spills, biodegradation and detoxification of oil-contaminated soils) [69].

6. Biostimulation and bioaugmentation

Biostimulation and bioaugmentation are the two most developed approaches among the bioremediation techniques. Their main purposes are the reduction of bioremediation time and the achievement of a complete removal of contaminant [4].

In biostimulation, nutrients and electron exchangers are injected into the contaminated site in order to stimulate the degrading ability of indigenous microorganisms [72]. As regards lab scale tests, nutrients are generally added as inorganic salts and as defined chemical species, while at the field scale, the nutrients are frequently added in the form of agro-wastes, organic wastes or inorganic fertilizers [63]. The main inorganic nutrients, usually added, are nitrogen and phosphorous, because the presence of organic toxic chemicals frequently induces an imbalance in the C:N:P ratio [73]. The main advantages of biostimulation approach are the low cost and the exploitation of indigenous microorganisms without the necessity of adaptation required by allochthonous species.

In bioaugmentation, allochthonous or enriched autochthonous microorganisms, able to metabolize a specific contaminant, are introduced in soil. In both cases, the homogeneous dispersion of the added biomass and its proliferation, in competition with native microorganisms, are the great challenges [63]. Moreover, bioaugmentation and biostimulation could be also coupled in order to further stimulate introduced biomass [74].

In fungal augmentation, high-quality inocula with high potentiality are necessary; consequently, specific methods have been developed for the production of fungal inocula. These inocula can be in the form of pelleted solid substrates, colonized by fungal mycelium, prepared from cheap agricultural and industrial by-products [4, 75]. Pelleted fungal inocula can be optimized in substrate composition to enhance fungal growth, degradation abilities and competitiveness against autochthonous soil microorganisms.

The bioaugmentation with autochthonous filamentous fungi for the cleanup of a historically contaminated site has been shown to be a successful bioremediation approach as described by [76]. These fungi were able to grow under nonsterile conditions and to degrade various aromatic hydrocarbons in the same contaminated soil.

In a recent review [77], the role of saprotrophic fungi in the biodegradation of xenobiotics and toxic metals in co-contaminated sites has been discussed along with the metabolic interactions between fungi and bacteria in a microbial consortium. Considering the occurrence of a mixed organic-inorganic contamination in brown field sites, the bioremediation mechanisms for combined pollution of PAHs and toxic metals by fungi and bacteria are also well documented [78].

7. Microcosm study at the lab scale

Microcosm studies are needed, before the in-field treatment, to evaluate microbial potential to degrade soil pollutants, the activity of the indigenous biomass and the most effective bioremediation strategy (i.e. biostimulation and/or bioaugmentation). In order to obtain information on the contaminant biodegradation in soil, the use of microcosms is a better approach than other kinds of laboratory tests [79]. Even if trials carried out at the lab scale do not always guarantee reproducible results on-site, due to chemical, physical and biological factors, they allow to verify the biodegradability of a certain compound. Hereafter, some of the most significant soil microcosm studies with fungi are reported.

One of the first studies, about PAH degradation in soil microcosm, was carried out with *P. chrysosporium* and *T. harzianum*, grown on wheat straw and then inoculated in naphthalene-contaminated soil. The biodegradation behaviour was monitored by means of naphthalene concentration measurement, CO_2 evolution as well as phytotoxicity tests [80].

Phanerochaete velutina and many litter-decomposing fungi (LDF) are potential degraders of soil organic matter. In the work of [81], they showed good growth, respiratory activity and MnP production on pine bark as co-substrate in microcosm. In the work of [82], the addition of *P. velutina*, cultivated on pine bark, to a PAH-contaminated soil was evaluated in microcosm and at the field scale. In the microcosm treatment (about 1 kg of soil), the bioaugmentation with fungi showed a positive effect on the biodegradation of the high molecular weight PAHs. On the contrary, in the field-scale experiment (about 2 tons of soil), carried out at lower starting concentration of PAHs, the degradation percentage (%) was similar in both the inoculated and non-inoculated soils.

The bioremediation of an aged PAH-contaminated soil in microcosm was demonstrated for an isolate of *Trichoderma reesei* [83]. The fungus metabolized benzo[a]pyrene in the presence of glucose as a co-metabolic substrate.

An isolate of *Chaetomium aureum* was able to halve the free Pb concentration in soil in about 2 months, irrespective of its association with indigenous microorganisms when inoculated in Pb-contaminated soil microcosms [84].

A microcosm study was conducted to optimize the degradation of weathered total petroleum hydrocarbons (TPH) in arid soils contaminated for more than a decade. Among fungi, *Aspergillus*, *Acremonium*, *Cryptococcus*, *Geotrichum* and *Penicillium* were the most widespread in these soils [85].

Different fungal strains (*Aspergillus*, *Fusarium*, *Rhizomucor* and *Emericella* spp.), isolated from a higher As contaminated agricultural soil, showed different detoxification mechanisms (biosorption/bioaccumulation and biovolatilization). They were able to reduce As contamination under in situ conditions as reported by [86].

In a study on bioremediation of petroleum hydrocarbons, a periodic biostimulation and bioaugmentation (PBB), by a single strain or a fungal consortium, was reported as the best biodegradation strategy [87]. PBB maintained the enzymatic activities of a fungal co-culture (*Pestalotiopsis* sp., *Polyporus* sp. and *Trametes hirsuta*) and enhanced the biodegradation rate, in particular during the early stage of remediation [73].

The biodegradation activity of *Byssochlamys nivea* and *Scopulariopsis brumptii* was evaluated in agricultural soil microcosms contaminated with pentachlorophenol (PCP), added with solid urban waste compost [88]. A synergistic effect of compost and fungal strains was observed with a reduction of more than 95% of PCP after 28 days of incubation. The detoxification role of the two fungal strains in the contaminated soil was also confirmed by toxicity assays [89].

8. Mushrooms as an emerging issue in mycoremediation

Mushroom application in the bioremediation field could be considered as an emerging technology; nevertheless, a lot of scientific works have appeared in the last years.

The biodegradation potential of mushroom species in soil has been reviewed by [9]. In this chapter, the mycelial capability of hyperaccumulate chemical elements, in particular heavy metals and radionuclides, along with the nutritional potential hazards due to mushroom consumption has been extensively discussed.

The biodegradation of recalcitrant pollutants like PAHs by WRF, the bioremediation of soil contaminated with engine oil by *Lentinus squarrosulus* and the decontamination of soils polluted with cement and battery wastes using *Pleurotus pulmonarius* were also reported by [29].

Many works on the edible mushroom *P. ostreatus* have been published. The biodegradation of the carcinogenic secondary metabolite aflatoxin B_1 (AFB_1), produced by *Aspergillus flavus* on rice straw [90] and on maize [91], was reported for this species. The mycoremediation of heavy metal-contaminated soils by means of different *Pleurotus* species was also reviewed in the work of [92]. In general, *Pleurotus* spp. are reported to be able to accumulate high levels of heavy metals; each species is characterized by different sensitivities towards the different metals and their concentration.

In the review of [10], mushroom bioaccumulation of different potentially toxic trace elements (PTEs) in the fruiting bodies was reported for *Phellinus badius*, *Amanita spissa*, *Lactarius piperatus*, *Suillus grevillei*, *Agaricus bisporus*, *Tricholoma terreum* and *Fomes fomentarius*. The accumulation capability was higher than that of plants, vegetables and fruits.

The bioremediation of crude oil-contaminated soil by an unidentified *Agaricomycetes* was demonstrated in the work of [93]. The addition of 10% of spent mushroom compost (SMC) allowed to degrade petroleum hydrocarbons over a short period of time.

9. Conclusion

The capability of micro- and macrofungi to degrade organic pollutants and to decrease heavy metal concentration in soil is a matter of fact. The growth morphology in soil (i.e. extended hyphal network), the low specificity of extracellular enzymatic complexes and the possibility to use toxic compounds as the growth substrate make filamentous fungi more advantageous in bioremediation processes when compared to other microorganisms. However, in the design of a soil mycoremediation process, some important aspects have to be considered such as the choice of the appropriate fungal strain and the evaluation of its possible interaction with the contaminated soil microbiota. To this end, microcosm studies represent a useful and simple method which allows to evaluate the feasibility of a biodegradation process.

Author details

Francesca Bosco* and Chiara Mollea
Politecnico di Torino, Torino, Italy

*Address all correspondence to: francesca.bosco@polito.it

References

[1] JRC. The state of soil in Europe. In: EEA Environment State and Outlook Report—SOER 2010. Luxembourg: Publications Office of the European Union; 2012. DOI: 10.2788/77361

[2] Panagos P, Liedekerke MV, Yigini Y, Montanarella L. Contaminated sites in Europe: Review of the current situation based on data collected through a European network. Journal of Environmental and Public Health. 2013;**2013**:1-11. DOI: 10.1155/2013/158764

[3] Schippers A, Glombitza F, Sand W, editors. Geobiotechnology. II. Energy Resources, Subsurface Technologies, Organic Pollutants and Mining Legal Principles. 1st ed. Berlin/Heidelberg: Springer-Verlag; 2014. DOI: 10.1007/978-3-662-44474-0. 200p

[4] Singh H, editor. Mycoremediation Fungal Bioremediation. Hoboken, New Jersey: John Wiley & Sons; 2006. DOI: 10.1002/0470050594. 592p

[5] Goltapeh EM, Danesh YR, Varma A, editors. Fungi as Bioremediators. 1st ed. Berlin/Heidelberg: Springer-Verlag; 2013. DOI: 10.1007/978-3-642-33811-3. 489p

[6] Baldrian P. Wood-inhabiting ligninolytic basidiomycetes in soils: Ecology and constraints for applicability in bioremediation. Fungal Ecology. 2008;**1**:4-12. DOI: 10.1016/j.funeco.2008.02.001

[7] Spina F, Cecchi G, Landinez-Torres A, Pecoraro L, Russo F, Wu B, et al. Fungi as a toolbox for sustainable bioremediation of pesticides in soil and water. Plant Biosystems. 2018;**152**(3):474-488. DOI: 10.1080/11263504.2018.1445130

[8] Deshmukh R, Khardenavis AA, Purohit HJ. Diverse metabolic capacities of fungi for bioremediation. Indian Journal of Microbiology. 2016;**56**(3):247-264. DOI: 10.1007/s12088-016-0584-6

[9] Chatterjee S, Sarma MK, Deb U, Steinhauser G, Walther C, Gupta DK. Mushrooms: From nutrition to mycoremediation. Environmental Science and Pollution Research. 2017;**24**:19480-19493. DOI: 10.1007/s11356-017-9826-3

[10] Ali A, GuoO D, Mahar A, Wang P, Shen F, Li R, et al. Mycoremediation of potentially toxic trace elements—A biological tool for soil cleanup: A review. Pedosphere. 2017;**27**(2):205-222. DOI: 10.1016/S1002-0160(17)60311-4

[11] Purnomo AS, Mori T, Putra SR, Kondo R. Biotransformation of heptachlor and heptachlor epoxide by white-rot fungus *Pleurotus ostreatus*. International Biodeterioration and Biodegradation. 2013;**82**:40-44. DOI: 10.1016/j.ibiod.2013.02.013

[12] Kulshreshtha S, Mathur N, Bhatnagar P. Mycoremediation of paper, pulp and cardboard industrial wastes and pollutants. In: Goltapeh E, Danesh Y, Varma A, editors. Fungi as Bioremediators. 1st ed. Berlin/Heidelberg: Springer; 2013. pp. 77-116. DOI: 10.1007/978-3-642-33811-3_4

[13] Joutey NT, Bahafid W, Sayel H, El Ghachtouli N. Biodegradation: Involved microorganisms and genetically engineered microorganisms. In: Chamy R, Rosenkranz F, editors. Biodegradation. London, UK: IntechOpen Limited; 2013. p. 290-320. DOI: 10.5772/56194

[14] Bharath Y, Singh SN, Keerthiga G, Prabhakar R. Mycoremediation of contaminated soil in MSW sites. In: Ghosh SK, editor. Waste Management and Resource Efficiency. Singapore: Springer Nature; 2019. pp. 321-329. DOI: 10.1007/978-981-10-7290-1_28

[15] de Lima Souza HM, Sette LD, da Mota AJ, do Nascimento Neto JF, Rodrigues A, de Oliveira TB, et al. Filamentous fungi isolates of contaminated sediment in the Amazon region with the potential for benzo(a)pyrene degradation. Water, Air, and Soil Pollution. 2016;**227**:431. DOI: 10.1007/s11270-016-3101-y

[16] Harms H, Schlosser D, Wick LY. Untapped potential: Exploiting fungi in bioremediation of hazardous chemicals. Nature Reviews. Microbiology. 2011;**9**:177-192. DOI: 10.1038/nrmicro2519

[17] Treu R, Falandysz J. Mycoremediation of hydrocarbons with basidiomycetes—A review. Journal of Environmental Science and Health. Part. B. 2017;**52**(3):148-155. DOI: 10.1080/03601234.2017.1261536

[18] Ellouze M, Sayadi S. White-rot fungi and their enzymes as a biotechnological tool for Xenobiotic bioremediation. In: El-Din M, Saleh H, editors. Management of Hazardous Wastes. London, UK: IntechOpen Limited; 2016. pp. 103-120. DOI: 10.5772/64145

[19] Abdel-Hamid AM, Solbiati JO, Cann IKO. Chapter one: Insights into lignin degradation and its potential industrial applications. In: Sariaslani S, Gadd GM, editors. Advances in Applied Microbiology. San Diego, CA: Academic Press/ Elsevier; 2013. pp. 1-28. DOI: 10.1016/ B978-0-12-407679-2.00001-6

[20] Voběrková S, Solčány V, Vršanská M, Vojtěch A. Immobilization of ligninolytic enzymes from white-rot fungi in cross-linked aggregates. Chemosphere. 2018;**202**:694-707. DOI: 10.1016/j.chemosphere.2018.03.088

[21] Manavalan T, Manavalan A, Heese K. Characterization of lignocellulolytic enzymes from white-rot fungi. Current Microbiology. 2015;**70**:485-498. DOI: 10.1007/s00284-014-0743-0

[22] Bumpus JA, Tien M, Wright D, Aust SD. Oxidation of persistent environmental pollutants by a white rot fungus. Science, New Series. 1985;**228**(4706):1434-1436

[23] Castro Faria d, AB. Evaluation of ectomycorrhizal respiration for remediation of pesticides in forestry. Advances in Plants & Agriculture Research. 2018;**8**(6):396-398. DOI: 10.15406/apar.2018.08.00357

[24] Meharg AA, Cairney JWG. Ectomycorrhizas—Extending the capabilities of rhizosphere remediation? Soil Biology and Biochemistry. 2000;**32**:1475-1484

[25] Rúa MA, Moore B, Hergott N, Van L, Jackson CR, Hoeksema JD. Ectomycorrhizal fungal communities and enzymatic activities vary across an ecotone between a forest and field. Journal of Fungi. 2015;**1**:185-210. DOI: 10.3390/jof1020185

[26] Kumar J, Atri NS. Studies on ectomycorrhiza: An appraisal. The Botanical Review. 2018;**84**:108-155. DOI: 10.1007/s12229-017-9196-z

[27] Li X, Wang Y, Pan Y, Yu H, Zhang X, Shen Y, et al. Mechanisms of Cd and Cr removal and tolerance by macrofungus *Pleurotus ostreatus* HAU-2. Journal of Hazardous Materials. 2017;**330**:1-8. DOI: 10.1016/j.jhazmat.2017.01.047

[28] Ji L, Zhang W, Yu D, Cao Y, Xu H. Effect of heavy metal-solubilizing microorganisms on zinc and cadmium extractions from heavy metal contaminated soil with *Tricholoma lobynsis*. World Journal of Microbiology and Biotechnology. 2012;**28**:293-301. DOI: 10.1007/s11274-011-0819-y

[29] Adenipekun CO, Lawal R. Uses of mushrooms in bioremediation: A review. Biotechnology and Molecular Biology Reviews. 2012;7(3):62-68. DOI: 10.5897/BMBR12.006

[30] Boonsong BS, Klaypradit W, Wilaipun P. Antioxidant activities of extracts from five edible mushrooms using different extractants. Agriculture and Natural Resources. 2016;**50**(2): 89-97. DOI: 10.1016/j.anres.2015.07.002

[31] Thakur M. Mushrooms as a biological tool in mycoremediation of polluted soils. In: Jindal T, editor. Emerging Issues in Ecology and Environmental Science. Cham, Switzerland: Springer Nature; 2019. pp. 27-42. DOI: 10.1007/978-3-319-99398-0_3

[32] He Z, Shentu J, Yang X, Baligar VC, Zhang T, Stoffella PJ. Heavy metal contamination of soils: Sources, indicators, and assessment. Journal of Ecological Indicators. 2015;**9**:17-18

[33] Lukić B, Panico A, Huguenot D, Fabbricino M, van Hullebusch ED, Esposito G. A review on the efficiency of landfarming integrated with composting as a soil remediation treatment. Environmental Technology Reviews. 2017;**6**(1):94-116. DOI: 10.1080/21622515.2017.1310310

[34] Pachana K, Wattanakornsiri A, Nanuam J. Heavy metal transport and fate in the environmental compartments. Naresuan University Science Journal. 2010;**7**(1):1-11

[35] Raj DD, Mohan B, Vidya Shetty BM. Mushrooms in the remediation of heavy metals from soil. International Journal of Environmental Pollution Control and Management. 2011;**3**:89-101

[36] IARC. Chemical Agents and Related Occupations Volume 100 F. A Review of Human Carcinogens. Lyon Cedex 08, France: Publication of International Agency for Research on Cancer; 2012

[37] Zhang JH, Xue Q, Gao H, Ma X, Wang P. Degradation of crude oil by fungal enzyme preparations from *Aspergillus* spp. for potential use in enhanced oil recovery. Journal of Chemical Technology and Biotechnology. 2016;**91**:865-875. DOI: 10.1002/jctb.4650

[38] Ameen F, Moslem M, Hadi S, Al-Sabri AE. Biodegradation of diesel fuel hydrocarbons by mangrove fungi from Red Sea Coast of Saudi Arabia. Saudi Journal of Biological Sciences. 2016;**23**: 211-218. DOI: 10.1016/j.sjbs.2015.04.005

[39] Dawoodi V, Madani M, Tahmourespour A, Golshani Z. The study of heterotrophic and crude oil-utilizing soil fungi in crude oil contaminated regions. Journal of Bioremediation & Biodegradation. 2015;**6**(2):1-5. DOI: 10.4172/2155-6199.1000270

[40] Abdel-Shafy HI, Mansour MSM. A review on polycyclic aromatic hydrocarbons: Source, environmental impact, effect on human health and remediation. Egyptian Journal of Petroleum. 2016;**25**:107-123. DOI: 10.1016/j.ejpe.2015.03.011

[41] Aranda E. Promising approaches towards biotransformation of polycyclic aromatic hydrocarbons with Ascomycota fungi. Current Opinion in Biotechnology. 2016;**38**:1-8. DOI: 10.1016/j.copbio.2015.12.002

[42] Aydin S, Karacay HA, Shahi A, Gokce S, Ince B, Ince O. Aerobic and anaerobic fungal metabolism and Omics insights for increasing polycyclic aromatic hydrocarbons biodegradation. Fungal Biology Reviews. 2017;**31**:61-72. DOI: 10.1016/j.fbr.2016.12.001

[43] Godambe T, Fulekar M. Bioremediation of petrochemical hydrocarbons (BTEX)—Review. Journal of Environmental Science and Pollution Research. 2017;**3**(3):189-199

[44] Min K, Freeman C, Kang H, Choi SU. The regulation by phenolic compounds of soil organic matter

dynamics under a changing environment. BioMed Research International. 2015;**2015**:1-11. DOI: 10.1155/2015/825098

[45] Kahru A, Pollumaa L, Reiman R, Ratsep A, Liiders M, Maloveryan A. The toxicity and biodegradability of eight main phenolic compounds characteristic to the oil-shale industry wastewaters: A test battery approach. Environmental Toxicology. 2000;**15**(5):431-442. DOI: 10.1002/1522-7278(2000)15:5<431::AID-TOX11>3.0.CO;2-T

[46] Kues U. Fungal enzymes for environmental management. Current Opinion in Biotechnology. 2015;**33**:268-278. DOI: 10.1016/j.copbio.2015.03.006

[47] Sinsabaugh S. Phenol oxidase, peroxidase and organic matter dynamics of soil. Soil Biology and Biochemistry. 2010;**42**:391-404. DOI: 10.1016/j.soilbio.2009.10.014

[48] Martínkova L, Kotik M, Markova E, Homolka L. Biodegradation of phenolic compounds by Basidiomycota and its phenol oxidases: A review. Chemosphere. 2016;**149**:373-382. DOI: 10.1016/j.chemosphere.2016.01.022

[49] Farrington JW. Chlorinated hydrocarbons. In: Thorpe SA, Turekian KK, editors. Encyclopedia of Ocean Sciences, Vol. 1., 1st ed. Amsterdam: Elsevier Academic Press; 2001. pp. 551-562. DOI: 10.1016/B978-0-12-409548-9.09090-4

[50] Marco-Urrea E, Garcia-Romera I, Aranda E. Potential of non-ligninolytic fungi in bioremediation of chlorinated and polycyclic aromatic hydrocarbons. New Biotechnology. 2015;**32**(6):620-628. DOI: 10.1016/j.nbt.2015.01.005

[51] Duarte RMBO, Matos JTV, Senesi N. Organic pollutants in soils. In: Duarte AC, Cachada A, Rocha-Santos T, editors. Soil Pollution From Monitoring to Remediation. London, UK: Elsevier/Academic Press; 2018. pp. 103-126. DOI: 10.1016/B978-0-12-849873-6.00005-4

[52] Dodgen LK, Li j WX, Lu Z, Gan JJ. Transformation and removal pathways of four common PPCP/EDCs in soil. Environmental Pollution. 2014;**193**:29-36. DOI: 10.1016/j.envpol.2014.06.002

[53] Ying GG, Kookana RS. Sorption and degradation of estrogen-like-endocrine disrupting chemicals in soil. Environmental Toxicology and Chemistry. 2005;**24**(10):2640-2645

[54] Chen F, Ying GG, Yang GF, Zhao JL, Wang L. Rapid resolution liquid chromatography-tandem mass spectrometry method for the determination of endocrine disrupting chemicals (EDCs), pharmaceuticals and personal care products (PPCPs) in wastewater irrigated soils. Journal of Environmental Science and Health, Part B. 2010;**45**(7):682-693. DOI: 10.1080/03601234.2010.502446

[55] Cajthaml T. Biodegradation of endocrine-disrupting compounds by ligninolytic fungi: Mechanisms involved in the degradation. Environmental Microbiology. 2015;**17**(12):4822-4834. DOI: 10.1111/1462-2920.12460

[56] Rodarte-Morales AI, Feijoo G, Moreira MT, Lema JM. Degradation of selected pharmaceutical and personal care products (PPCPs) by white-rot fungi. World Journal of Microbiology and Biotechnology. 2011;**27**:1839-1846. DOI: 10.1007/s11274-010-0642-x

[57] Levasseur A, Lomascolo A, Chabrol O, Ruiz-Duenas FJ, Boukhris-Uzan E, Piumi F, et al. The genome of the white-rot fungus *Pycnoporus cinnabarinus*: A basidiomycete model with a versatile arsenal for lignocellulosic biomass breakdown.

BMC Genomics. 2014;**15**:486. DOI: doi.org/10.1186/1471-2164-15-486

[58] Senthivelan T, Kanagaraj J, Panda RC. Recent trends in fungal laccase for various industrial applications: An eco-friendly approach—A review. Biotechnology and Bioprocess Engineering. 2016;**21**:19-38. DOI: 10.1007/s12257-015-0278-7

[59] Hofrichter M, Ullrich R, Pecyna MJ, Liers C, Lundell T. New and classic families of secreted fungal heme peroxidases. Applied Microbiology and Biotechnology. 2010;**87**:871-897. DOI: 10.1007/s00253-010-2633-0

[60] Lauber C, Schwarz T, Khanh Nguyen Q, Lorenz P, Lochnit G, Zorn H. Identification, heterologous expression and characterization of a dye-decolorizing peroxidase of *Pleurotus sapidus*. AMB Express. 2017;**7**(164):1-15. DOI: 10.1186/s13568-017-0463-5

[61] Mäkelä MR, Marinović M, Nousiainen P, Liwanag AJ, Benoit I, Sipilä J, et al. Aromatic metabolism of filamentous fungi in relation to the presence of aromatic compounds in plant biomass. Advances in Applied Microbiology. 2015;**91**:63-137. DOI: 10.1016/bs.aambs.2014.12.001

[62] Morel M, Ngadin AA, Droux M, Jacquot JP, Gelhaye E. The fungal glutathione S-transferase system. Evidence of new classes in the wood-degrading basidiomycete *Phanerochaete chrysosporium*. Cellular and Molecular Life Sciences. 2009;**66**:3711-3725. DOI: 10.1007/s00018-009-0104-5

[63] Adams GO, Fufeyin PT, Okoro SE, Ehinomen I. Bioremediation, biostimulation and bioaugmention: A review. International Journal of Environmental Bioremediation & Biodegradation. 2015;**3**(1):28-39. DOI: 10.12691/ijebb-3-1-5

[64] Chanda A, Gummadidala PM, Goma OM. Mycoremediation with mycotoxin producers: A critical perspective. Applied Microbiology and Biotechnology. 2016;**100**:17-29. DOI: 10.1007/s00253-015-7032-0

[65] Agunbiade FO, Moodle B. Pharmaceuticals as emerging organic contaminants in Umgeni River water system, KwaZulu-Natal, South Africa. Environmental Monitoring and Assessment. 2014;**186**:7273-7291. DOI: 1.1007/s10661-014-3926-z

[66] Esteban S, Gorga M, Petrovic M, González-Alonso S, Barceló D, Valcárcel Y. Analysis and occurrence of endocrine-disrupting compounds and estrogenic activity in the surface waters of Central Spain. Science of the Total Environment. 2014;466-467. DOI: 10.1016/j.scitotenv.2013.07.101

[67] Leonardi V, Sasek V, Petruccioli M, D'Annibale A, Erbanova P, Cajthaml T. Bioavailability modification and fungal biodegradation of PAHs in aged industrial soils. International Biodeterioration and Biodegradation. 2007;**60**:165-170. DOI: 10.1016/j.ibiod.2007.02.004

[68] Prakash V. Mycoremediation of environmental pollutants. International Journal of ChemTech Research. 2017;**10**(3):149-155

[69] Shekhar S, Sundaramanickam A, Balasubramanian T. Biosurfactant producing microbes and their potential applications: A review. Critical Reviews in Environmental Science and Technology. 2015;**45**(14):1522-1554. DOI: 10.1080/10643389.2014.955631

[70] Gao Y, Li Q, Ling W, Zhu X. Arbuscular mycorrhizal phytoremediation of soils contaminated with phenanthrene and pyrene. Journal of Hazardous Materials. 2011;**185**:703-709. DOI: 10.1016/j.jhazmat.2010.09.076

[71] Lin W, Brauers G, Ebel R, Wray V, Berg A, Proksch P. Novel chromone derivatives from the fungus Aspergillus versicolor isolated from the marine sponge *Xestospongia exigua*. Journal of Natural Products. 2003;**66**(1):57-61. DOI: 10.1021/np020196b

[72] Sayara T, Borràs E, Caminal G, Sarrà M, Sánchez A. Bioremediation of PAHs-contaminated soil through composting: Influence of bioaugmentation and biostimulation on contaminant biodegradation. International Biodeterioration and Biodegradation. 2011;**65**(6):859-865. DOI: 10.1016/j.ibiod.2011.05.006

[73] Wu M, Dick WA, Li W, Wang X, Yang Q, Wang T, et al. Bioaugmentation and biostimulation of hydrocarbon degradation and the microbial community in a petroleum-contaminated soil. International Biodeterioration and Biodegradation. 2016;**107**:158-164. DOI: :10.1016/j.ibiod.2015.11.019

[74] Ghaly AE, Yusran A, Dave D. Effects of biostimulation and bioaugmentation on the degradation of pyrene in soil. Journal of Bioremediation & Biodegradation. 2013;S7(005):1-13. DOI: 10.4172/2155-6199.S7-005

[75] Elgueta S, Santos C, Lima N, Diez MC. Immobilization of the white-rot fungus *Anthracophyllum discolor* to degrade the herbicide atrazine. AMB Express. 2016;**6**(104):1-11. DOI: 10.1186/s13568-016-0275-z

[76] D'Annibale A, Rosetto F, Leonardi V, Federici F, Petruccioli M. Role of autochthonous filamentous fungi in bioremediation of a soil historically contaminated with aromatic hydrocarbons. Applied and Environmental Microbiology. 2006;**72**(1):28-36. DOI: 10.1128/AEM.72.1.28-36.2006

[77] Ceci A, Pinzari F, Russo F, Persiani AM, Gadd GM. Roles of saprotrophic fungi in biodegradation or transformation of organic and inorganic pollutants in co-contaminated sites. Applied Microbiology and Biotechnology. 2019;**103**:53-68. DOI: 10.1007/s00253-018-9451-1

[78] Liu SH, Zeng GM, Niu QY, Liu Y, Zhou L, Jiang LH, et al. Bioremediation mechanisms of combined pollution of PAHs and heavy metals by bacteria and fungi: A mini review. Bioresource Technology. 2017;**224**:25:33. DOI: 10.1016/j.biortech.2016.11.095

[79] Caracciolo AB, Bottoni P, Grenni P. Microcosm studies to evaluate microbial potential to degrade pollutants in soil and water ecosystems. Microchemical Journal. 2013;**107**:126-130

[80] Mollea C, Bosco F, Ruggeri B. Fungal biodegradation of naphthalene: Microcosms studies. Chemosphere. 2005;**60**:636-643. DOI: 10.1016/j.chemosphere.2005.01.034

[81] Valentìn L, Kluczek-Turpeinen B, Oivanen P, Hatakka A, Steffen K, Tuomela M. Evaluation of basidiomycetous fungi for pretreatment of contaminated soil. Journal of Chemical Technology and Biotechnology. 2009;**84**:851-858

[82] Winquist E, Björklöf K, Schultz E, Räsänen M, Salonen K, Anasonye F, et al. Bioremediation of PAH-contaminated soil with fungi—From laboratory to field scale. International Biodeterioration and Biodegradation. 2014;**6**:238-247

[83] Yao L, Teng Y, Luo Y, Christie P, Ma W, Liu F, et al. Biodegradation of polycyclic aromatic hydrocarbons (PAHs) by *Trichoderma reesei* FS10-C and effect of bioaugmentation on an aged PAH contaminated soil. Bioremediation Journal. 2015;**19**:9-17

[84] Da Silva Júnior FMR, Martins Volcão L, Coutelle Hoscha L, Pereira SV.

Growth of the fungus *Chaetomium aureum* in the presence of lead: Implications in bioremediation. Environment and Earth Science. 2018;77:275

[85] Ramadass K, Smith E, Palanisami T, Mathieson G, Srivastava P, Megharaj M, et al. Evaluation of constraints in bioremediation of weathered hydrocarbon-contaminated arid soils through microcosm biopile study. International Journal of Environmental Science and Technology. 2015;**12**:3597-3612

[86] Singh M, Srivastava PK, Verma PC, Kharwar RN, Singh N, Tripathi RD. Soil fungi for mycoremediation of arsenic pollution in agriculture soils. Journal of Applied Microbiology. 2015;**119**:1278-1290

[87] Yanto DHY, Hidayat A, Tachibana S. Periodical biostimulation with nutrient addition and bioaugmentation using mixed fungal cultures to maintain enzymatic oxidation during extended bioremediation of oily soil microcosms. International Biodeterioration and Biodegradation. 2017;**116**:112-123

[88] Bosso L, Scelza R, Testa A, Cristinzio G, Rao MR. Depletion of pentachlorophenol contamination in an agricultural soil treated with *Byssochlamys nivea*, *Scopulariopsis brumptii* and urban waste compost: A laboratory microcosm study. Water, Air, and Soil Pollution. 2015;**226**:183

[89] Hechmi N, Bosso L, El-Bassi L, Scelza R, Testa A, Jedidi N, et al. Depletion of pentachlorophenol in soil microcosms with *Byssochlamys nivea* and *Scopulariopsis brumptii* as detoxification agents. Chemosphere. 2016;**165**:547-554

[90] Das A, Bhattacharya S, Palaniswamy M, Angayarkanni J. Aflatoxin B1 degradation during co-cultivation of *Aspergillus flavus* and *Pleurotus ostreatus* strains on rice straw. Biotech. 2015;**5**:279-284

[91] Jackson LW, Pryor BM. Degradation of aflatoxin B1 from naturally contaminated maize using the edible fungus *Pleurotus ostreatus*. AMB Express. 2017;**7**:110

[92] Kapahi M, Sachdeva S. Mycoremediation potential of *Pleurotus* species for heavy metals: A review. Bioresources and Bioprocessing. 2017;**4**:32

[93] Mohammadi-Sichani MM, Mazaheri Assadi M, Farazmand A, Kianirad M, Ahadi AM, Hadian Ghahderijani H. Bioremediation of soil contaminated crude oil by *Agaricomycetes*. Journal of Environmental Health Science and Engineering. 2017;**15**:8

Chapter 11

Arsenic Phytoremediation: Finally a Feasible Approach in the Near Future

Cristian Mateo, Micaela Navarro, Cristina Navarro and Antonio Leyva

Abstract

Arsenic, a class-1 carcinogenic, is a ubiquitous metalloid found in the atmosphere, soils, natural waters, and organisms. The World Health Organization (WHO) estimates that hundred million people worldwide might be chronically exposed to arsenic in drinking water at concentrations above the safety standard. Conventionally applied techniques to remove arsenic species show low removal efficiency, high operational costs, and high-energy requirements. The biological methods, especially phytoremediation, could be cost-effective for protecting human health and the environment from toxic metal contamination. Plants, as sessile organisms, have developed an extraordinary capacity to tolerate arsenic through three main strategies: uptake repression, sequestration into the vacuole, or extrusion. Therefore, arsenic perception and tolerance require a coordinated response that involves arsenic transporters, extrusion pumps, vacuole transporters, and the activation of the phytochelatin biosynthetic pathway. For phytoremediation to become a feasible strategy for arsenic removal from contaminated sites, it is essential to completely understand the molecular mechanisms of arsenic uptake, extrusion, and sequestration, as well as how this response is coordinated. The new genome-wide technologies provide a unique opportunity to understand the molecular mechanisms underlying arsenic perception and accumulation in plants that will open up new possibilities for phytoremediation of arsenic-contaminated waters and soils.

Keywords: arsenic, phytoremediation, hyperaccumulator, biotechnological approaches, *Arabidopsis thaliana*, *Pteris vittata*, rice

1. Introduction

Arsenic is a poison naturally present in the Earth's crust, where it constitutes the 20th most abundant element. Contamination with this element is derived from natural activities, such as volcanism, erosion, or leaching into aquifers, as well as from anthropogenic activities, like mining, smelting ores, or industry. The most abundant forms of arsenic in the environment are the inorganic oxyanions of arsenite (As(III)) and arsenate (As(V)), both of them highly soluble in water [1–3]. As(III) is more toxic and relatively mobile in contaminated soils, whereas As(V) is considered relatively less toxic.

Arsenic from contaminated soil and water resources poses an environmental threat for all living beings, since it is bioavailable for crops and animals, eventually entering the food chain. Once inside the cell, As(V) interferes with phosphate metabolism, due to its structural analogy to this compound, while As(III) binds to sulfhydryl groups of proteins affecting their functions. Arsenic toxicity is, thus, mostly derived from its interference with enzymes involved in DNA synthesis and repair and cellular energy homeostasis, among others [4]. This metalloid is a well-documented genotoxic agent and class I carcinogen affecting the skin and internal organs. Arsenic effects on human health have been a matter of comprehensive reviews [5–10]. For these reasons, chronic arsenic exposure, reported in many regions of the world, constitutes an environmental and public health problem of devastating proportions, particularly in Bangladesh [11, 12]. Besides the exposure to contaminated water, another major source of chronic arsenic exposure is derived from dietary consumption of contaminated rice. This crop accumulates relatively high amounts of arsenic and constitutes the main staple food for over 3 billion people across the world, mostly in Asia, being also extensively used for infant feeding [13–17].

For all these reasons, it is imperative to develop strategies to efficiently limit the effects of arsenic contamination and its entry in the food chain. Physicochemical treatments have been assayed for arsenic-contaminated waters, and recently nanomaterials and metal-organic frameworks (MOFs) are emerging as new promising adsorbent materials. In general, physicochemical treatments are effective at high arsenic concentrations but fail to remove it when contamination levels are low. Another drawback of these techniques is their economic cost, hardly affordable in economically poor rural areas [18, 19].

An alternative approach is the use of living organisms to mitigate arsenic contamination, a strategy named bioremediation. This strategy takes advantage of the ability of microorganisms and plants to tolerate and accumulate arsenic present in nature [20]. The use of plants to clean up contaminated environments is an environmentally friendly green technology, well-accepted sociologically, relatively easy to implement, and potentially cost-effective that can be used to treat both contaminated soils and waters [21–23]. The main drawbacks of this strategy include the requirement of prolonged periods of time to be effective, lack of reproducibility due to environmental constraints, and the handling of the plant-fixed arsenic waste disposal. Another limitation is that high concentrations of arsenic may inhibit plant growth and, therefore, phytoremediation performance. A recent report has shown that plants accumulating arsenic could efficiently extract and convert it into valuable compounds, such as arsenic nanoparticles, with potential medical applications [24]. In addition, recent advances in the understanding of arsenic toxicity and the pathways of arsenic uptake, accumulation, and tolerance in different plant species, mostly in *Arabidopsis thaliana* (*A. thaliana*) and rice, have enabled the identification of potential biotechnological targets to overcome these limitations and will improve efficiency of arsenic phytoremediation in the near future [25–28].

In this review, we focus on the phytoremediation strategy to combat arsenic poisoning: we provide a brief overview on the mechanisms of arsenic perception and signaling in different organisms, as well as a survey of plants with proven or potential use for arsenic phytoremediation. Finally, we address several biotechnological approaches that could potentially improve efficiency of arsenic phytoextraction and enable the development of safer edible crops.

2. Arsenic perception and signaling

In this section, we present an overview of the known molecular mechanisms involved in the arsenic response particularly in yeast and plants that will be the basis to design new applied strategies in the near future.

2.1 Arsenic response studies in prokaryotes: the pioneer studies of ars operon in *E. coli*

The identification of arsenic resistance genes in prokaryotes was first reported over 50 years ago [29]. This laboratory identified one strain of *Staphylococcus aureus* (*S. aureus*) resistant to antibiotic, arsenic and several heavy metal compounds, due to the presence of a resistance factor (R-factor) that was located at the plasmid, pI258. Later on, another R-factor, responsible for arsenic resistance, was found in *Escherichia coli* (*E. coli*), the R773 plasmid [30]. This was the very beginning of arsenic detoxification gene discoveries. Few years later, arsenic resistance genes were cloned from the *E. coli* R773 plasmid, confirming that this plasmid is responsible for arsenic tolerance and contains all the genes involved in the arsenic response, named the *ars* operon [31]. Since then, many works have been done to elucidate the complexity of the R-factors and the genes that constitute the *ars* operon in several species [32].

In this review, we will describe the main genes composing the *ars* operon and their function; we will not focus on the distribution of this operon in prokaryotes, comprehensively reviewed previously [32].

The *ars* operon is widely variable, comprising a group of genes represented in different species with a different range of complexity [32] responsible for *ars*enic tolerance in prokaryotes. For instance, the *S. aureus ars* operon, named *arsRBC*, is composed of three genes, *arsR*, *arsB*, and *arsC*. However, other species such *E. coli* contain additional genes in the operon, like *arsD* and *arsA*.

One of the most important components of the *ars* operon is *arsR*, the key regulator of arsenic response in prokaryotes. This gene encodes a metalloregulatory protein (ArsR) that acts as a repressor of the whole operon [33, 34]. ArsR is a member of the SmtB/ArsR family [35], a transcriptional repressor in its homodimeric form that contains a helix-turn-helix DNA-binding domain and metal-binding sites [35–38]. Under non-arsenic conditions, *arsR* directly binds to the *ars* operon repressing the expression of the downstream genes. In the presence of arsenic, even at low concentrations, this metalloid interacts with the ArsR homodimer, inducing its dissociation from the DNA [35]. Therefore, ArsR is considered the main sensor of the system through direct interaction with arsenic.

Another component of the operon, *arsC*, encodes an arsenate reductase [39] involved in the reduction of As(V) into As(III) using glutathione (GSH) and glutaredoxin (GRX) as electron donors [40]. On the other hand, different variants of ArsC from other species (*S. aureus*) use thioredoxin as an alternative electron donor [40]. The third gene of the operon, *arsB*, encodes a membrane protein which functions as a proton antiporter, involved in As(III) extrusion [41]. The versatility of ArsB is extraordinary. In many species, ArsB is present as a single component in which the As(III) extrusion is coupled to electrochemical energy, in an ATP-independent way [41]. In contrast, in many other species, such as *E. coli*, ArsB appears complexed with *arsA*. ArsA encodes an ATPase that forms a complex with ArsB that functions as an efficient ATPase-dependent arsenic pump, named ArsAB. ArsA displays a specific metal-binding domain in which As(III) is coupled. Several studies confirm that the presence of both genes gives an extraordinary bacterial survival advantage under arsenic stress conditions compared to the bacterial species holding only the ArsB protein.

The last component is *arsD*. Initially, ArsD was described as a metalloregulatory protein [42] with a proposed role in the transcriptional regulation of the ars operon, similar to ArsR. Few years later, it was shown that ArsD is a metallochaperone and not a regulatory protein [43]. This was an important contribution due to the fact that very little was known about metallochaperones by that time. The authors suggested that, under non-arsenic conditions, ArsD weakly interacts with the complex

ArsAB (the energetic arsenic pump). However in the presence of As(III), this chemical species is able to bind to ArsD [44], forming the complex ArsD-As(III), which can subsequently strongly interact with ArsAB, and then, As(III) is extruded. Hence, ArsD provides an efficient fine-tune mechanism of As(III) extrusion tightly modulated by As(III) concentration.

In conclusion, in non-arsenic conditions, ArsR stays attached to the operator of the operon and halts the transcription and expression *of the ars-resistant genes*. Conversely, when arsenic is incorporated into the cells through aquaporins or high-affinity phosphate transporters (for As(III) and As(V), respectively), ArsR directly interacts with the metalloid, triggering the operon transcription. ArsC reduces As(V) into As(III), which directly binds to ArsD, which in turn brings As(III) together with ArsAB, leading to the extrusion of the metalloid in an ATP-dependent manner. The detailed knowledge of all these mechanisms involved in arsenic tolerance paved the way to understand how eukaryotes are able to cope with the presence of arsenic.

2.2 *Saccharomyces cerevisiae*: a suitable model to study arsenic response in eukaryotes

Saccharomyces cerevisiae (*S. cerevisiae*) is an excellent model for the study of biological processes in eukaryotic organism, including arsenic stress responses. In this model, a single locus comprising three genes, *ACR1*, *ACR2*, and *ACR3*, is mainly in charge of the arsenic response and tolerance [45].

ACR1 encodes an AP-1-like transcription factor that has been recently characterized [46]. *ACR1*, also named Yap8, is able to interact directly with As(III) due to the presence of three cysteines, similar to ArsR in prokaryotes. When Yap8 interacts with the metalloid, it suffers a conformational change that avoids its ubiquitin-proteasome degradation. As a result, Yap8 protein remains stabilized bonded to the *ACR2/3* promoter (both genes are controlled by the same promoter, but they are transcribed in opposite directions). In the presence of As(III), Yap8 facilitates the recruitment of RNA polymerase II and triggers the transcription of *ACR2* and *ACR3*. *ACR2* encodes an arsenate reductase enzyme that reduces As(V) to As(III) [47]. However, previous experiments have demonstrated that ACR2 alone is not capable of increasing arsenic tolerance in yeast, since it requires the expression of an additional member of the cluster, ACR3. *ACR3* encodes a transmembrane protein directly involved in As(III) transport [48]. It has been proved that *acr3* mutants are sensitive to both As(V) and As(III), leading to the conclusion that this transporter is not only able to extrude As(III) to the media but also to incorporate it into cell compartments.

Similarly to ArsR in prokaryotes, Yap8 is considered the arsenic sensor in yeast. There is only one difference between the mechanisms of action of both transcription factors: while ArsR is a repressor, Yap8 acts as an activator. The identification of Yap8 implied an important contribution, since this protein was the first arsenic sensor described in eukaryotes, giving rise to a promising model on how other arsenic sensors may act in higher organisms including plants.

2.3 Arsenic response in eukaryotes: *Arabidopsis thaliana*

We will focus this section on the mechanisms of arsenic uptake and detoxification in the plant model *A. thaliana*. This basic knowledge is crucial to develop novel strategies for phytoremediation of contaminated soils and waters and to obtain staple crops able to produce safe food when grown in contaminated lands. In recent years, huge efforts have been made to understand the underlying molecular mechanisms of accumulation and tolerance of arsenic in plants [25–28]. Although much knowledge has been acquired over the last years, several aspects of the signal transduction pathway are still completely unknown; in particular, the desired arsenic sensor still needs to be identified in plants.

The mechanisms of incorporation and extrusion of As(V) and As(III) have been extensively described in *A. thaliana*. The prevalent chemical form of arsenic in soils is As(V); As(V) uptake is mediated by high-affinity phosphate transporters (PHT) from the *PHT1* family, mainly *PHT1;1* and *PHT1;4*, as a consequence of the structural similarity between these two anions [49]. As(III), present in paddy soils and anoxic environments, is incorporated by aquaporins from the nodulin 26-like intrinsic protein (NIP) subfamily, mainly *nip1;1* [50]. This family has been shown to be in charge of As(III) uptake and extrusion and in root-to-shoot translocation. Consistently, several studies confirm that *nip1;1* mutants are more tolerant to As(III) [50–52].

Once inside the cell, As(V) is rapidly transformed into As(III) through an As(V) reductase enzyme. Initially, AtACR2 was thought to be the main arsenate reductase in plants [53], which was identified by sequence homology with the yeast ACR2. A few years later, whether AtACR2 was the main As(V) reductase raised many questions among the scientific community [54], and a major role of AtACR2 remains to be confirmed. Later on, our group identified the gene (At2g21045) which encodes the major As(V) reductase in *A. thaliana*. The name of the gene refers to the QTL1, which confers the As(V) tolerance in *A. thaliana* as we reported by natural variation studies [55]. Afterwards the relevance of this protein in the arsenic response were corroborated by an independent group, which named it HAC1 from high arsenic content [56].

Once As(V) is reduced to As(III), it can be extruded by the previously mentioned NIP transporters. Even though As(V) tolerance in *nip1;1* mutants has not been tested yet, they are likely to display an As(V) hypersensitive phenotype, since they are not capable of extruding As(III) to the media. An alternative pathway for As(III) detoxification involves its sequestration by phytochelatins (PCs), forming PC-As(III) conjugates. PCs are peptides synthesized from glutathione and hold the basic structure (ɤ-Glu-Cys)n-Gly where n is a number ranging from 2 to 11 [57]. Indeed, PCs could be considered functional analogues of ArsD in terms of directly binding to As(III) [58]. ABCC transporters are ATP-dependent transporters responsible for the incorporation of PC-As(III) conjugates into the vacuole and thus essential for arsenic detoxification in *A. thaliana* [59].

In conclusion, arsenic in plants is incorporated into the cells either through the high-affinity phosphate transporters PHT (As(V)) or through NIPs (As(III)). Once inside the cell, it is reduced by ARQ1/HAC1 and subsequently gets sequestered by PCs. The resulting PCs-As(III) complexes interact with ABCC transporters mediating PCs-As(III) transport into the vacuoles. Alternatively, a fraction of As(III) can be extruded into the media through the NIPs transporters (**Figure 1**).

Overall, the detoxification mechanisms have been designed with the aim of protecting living organisms from the most dangerous chemical form of arsenic, As(III). Nevertheless, a novel phosphate vacuolar transporter has been recently identified [60]. Those mutants show an As(V) resistant phenotype since they cannot accumulate phosphate inside the vacuoles, and as a consequence it is likely that phosphate present in the cytoplasm provoked the repression of the As(V)/phosphate transporter *PHT1;1* preventing the entry of As(V). This novel mechanism provides the first insights into the crosstalk between As(V) and phosphate signal transduction pathways.

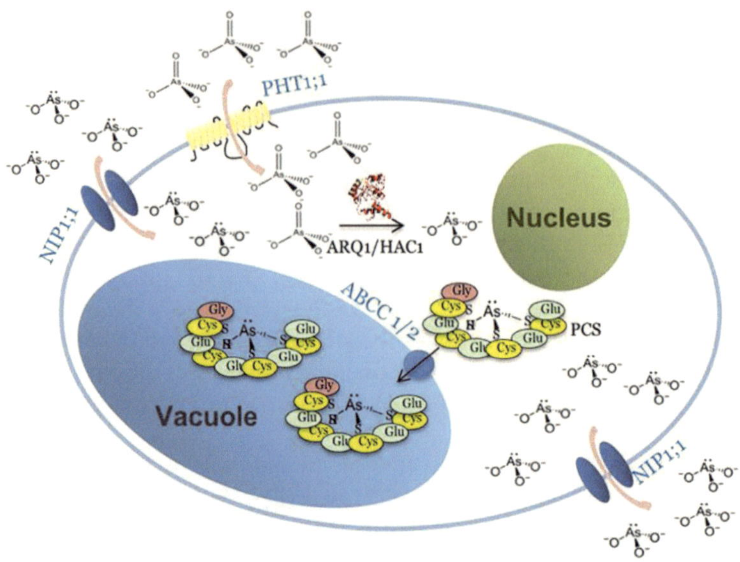

Figure 1.
Mechanisms of arsenic uptake and detoxification in A. thaliana.

3. Plant species with potential use for arsenic phytoremediation

The mechanisms of arsenic tolerance in plants—discussed above—confer adaptive responses that enable plant growth in the presence of this metalloid. There is an immense natural diversity in the arsenic response among different plant species. Some of them have developed an extraordinary capacity to grow in the presence of arsenic and hyperaccumulate this metalloid, holding a great potential to be used for phytoremediation strategies. The most critical parameter in plant species required for phytoremediation is a high extraction capacity. This can be achieved in different ways: via hyperaccumulation of the contaminant inside the cells or via high growth potential or high biomass production in the presence of the toxic compound. They should also be robust and adapted to a variety of biogeochemical environments, seasonal fluctuations, and climatic conditions, in order to obtain good extraction rates in different locations over the years. In addition, the ability to detoxify a variety of xenobiotics in complex degraded environments would be desirable.

A plant species is considered arsenic hyperaccumulator if it is able to accumulate more than 1000 $\mu g\ g^{-1}$ of dry weight [61, 62]. The brake fern *Pteris vittata* (*P. vittata*) is the arsenic hyperaccumulator species par excellence, being the most extensively reported and the best characterized until the moment [63–69]. The arsenic phytoremediation potential of *P. vittata* was first described in 2001. It was reported to tolerate as much as 1500 $\mu g\ g^{-1}$ arsenic in soil, accumulating up to 15,861 $\mu g\ g^{-1}$ in 2 weeks and up to 22,630 $\mu g\ g^{-1}$ in 6 weeks, under these conditions [70]. Since then, many reports have confirmed its arsenic hyperaccumulator capacity in greenhouse experiments and field conditions [63–69]. Some of them showed its utility to reduce arsenic contamination in crops grown intercropped with this fern [71–73]. Several studies have been conducted in order to elucidate the molecular mechanisms underlying the exceptional arsenic accumulation capacity and tolerance of *P. vittata*. Different aspects of arsenic uptake, detoxification, and tolerance have been characterized, showing differential strategies in relation to non-hyperaccumulator plants [74–82]. For example, the existence of an arsenic-inducible As(V)/phosphate transporter with high affinity for As(V) that could account for its great arsenic accumulation capacity has been recently described [74, 83]. Likewise, a differential antioxidant response could be the mechanism underlying the different degrees of tolerance observed among different species from the *Pteris* genus [80]. Therefore, most probably, the extraordinary extraction capacity and tolerance of *P. vittata* result from a specific combination of all these new mechanisms. Genome-wide studies are providing vast information that will help to have a more complete picture of these mechanisms in the near future [81–83]. Among them, a relevant recent report has shown the utility of transcriptomic and tonoplast-proteomic strategies to uncover a regulatory system mediated by alternative splicing and long noncoding RNAs in response to As(III) and As(V) treatments, as well as the identification of different transporter families and other genetic components underlying arsenic response and tolerance in this hyperaccumulator fern [82]. Besides its intrinsic role in phytoremediation as hyperaccumulator, the characterization of the molecular mechanisms operating in *P. vittata* will provide new regulatory elements that may confer increased arsenic accumulation capabilities in other species by genetic engineering.

Other species have been evaluated for their arsenic phytoremediation potential. Many of them are endemic species that have been selected because of their ability to grow in arsenic-contaminated areas, and thus they are good candidates for their use as arsenic phytoremediators [61, 63, 84–87]. As mentioned above, the remediation of arsenic-contaminated waters and paddy soils deserves special attention due to safety concerns. Several wetland plants and aquatic macrophytes have been shown to hyperaccumulate arsenic when growing in contaminated environments and could contribute to solving this devastating trouble [66, 88–94].

4. Potential transgenic/biotechnological approaches for phytoremediation and arsenic-free crop development

As mentioned above, huge efforts have been made in recent years in order to understand the underlying molecular mechanisms of accumulation and tolerance of arsenic in plants. These initiatives provide a basic knowledge that is crucial to develop novel phytoremediation strategies for environmental cleanup and for producing safe staple crops grown in contaminated lands. The key elements of the arsenic response in the model plant *A. thaliana* and other plant species will make possible to tailor the arsenic uptake and accumulation in crops for different purposes [28, 95, 96].

Phytoremediation of arsenic-contaminated areas requires the development of new plant varieties with enhanced uptake and accumulation of this metalloid in order to remove as much arsenic as possible from the environment. On the contrary, inhibiting the arsenic uptake and translocation to the edible parts and promoting its extrusion outside the cells will be imperative for the development of safe and productive crops grown on arsenic-contaminated lands [25, 96].

4.1 Generation of arsenic transgenic plants for phytoremediation purposes

One of the drawbacks of the phytoremediation approach is that it may have low efficiency as a consequence of phytotoxicity when plants are exposed to high arsenic levels. Moreover, some natural hyperaccumulators do not produce enough biomass, which is crucial for successful phytoremediation, and are also restricted to very specific climatic conditions [96]. Next-generation "omic" approaches are paving the way to increase plant tolerance and extraction of arsenic, holding promising results for phytoremediation [97]. Arsenic uptake, accumulation, and tolerance can be augmented through the modulation of influx/efflux plasma membrane transporters, the regulation of the arsenate reductase activity, and the increase of the amount of PCs and glutaredoxins [98]. The coordination of this element in the arsenic perception and tolerance is a key aspect to sequester arsenic from the cytosol into the vacuoles or its translocation from root to shoot via xylem loading [96, 99].

4.1.1 Arsenic uptake enhancement: phytoextraction

The knowledge acquired in studying the mechanisms of arsenic uptake and translocation by hyperaccumulators such as *P. vittata* (see above) provides important information to generate promising phytoremediation tools. Root-to-shoot translocation studies will contribute to obtain plants with increased shoot arsenic content, which could greatly benefit phytoremediation applications [98].

P. vittata displays an enhanced As(V) uptake derived from both increased *PvPHT1;3* expression and high affinity of this Pi/As(V) transporter for As(V). This transporter has greater affinity for As(V) than its counterpart from *A. thaliana*, *AtPht1;5* [74]. Therefore, we propose the heterologous expression of the transporter *PvPHT1;3*, which is more specific for As(V) uptake over Pi, in a high biomass plant as a potential strategy for extracting large amount of arsenic from contaminated lands or waters.

Another interesting example of a biotechnological application has been provided by the overexpression of the *PvACR3* transporter from *P. vittata* in *A. thaliana* [78]. This transporter mediates the translocation and storage of As(III) into the vacuolar system in *P. vittata*. *A. thaliana*-overexpressing plants display a significant increase of As(III) export from root to shoot, as well as an increment of arsenic tolerance. The strong expression of this transporter makes it to localize in the plasma membrane of the transgenic plants. This was expected to increase As(III) extrusion to the external medium or its sequestration in the root vacuoles, but there was an enhanced translocation of As(III) into the aerial parts instead [100]. Thus, it seems to be useful to constitutively express *PvACR3* transporter in a vigorous crop to facilitate root-to shoot translocation for phytoremediation purposes.

Importantly, the next-generation gene-editing CRISPR/Cas9 technology is nowadays an emerging tool to obtain improved crops [101]. This technology is target-specific and allows targeting multiple genes in the genome with high efficiency and specificity. Thereby, this system opens up the possibility to obtain precisely edited crops with enhanced arsenic extraction and accumulation. For example, engineering the aquatic plant *Lemna* with CRISPR/Cas9 for point mutations in the As(V)/phosphate transporters and As(III)-PCs vacuolar transporters at the same time would be a feasible strategy for cleaning arsenic-contaminated waters.

4.1.2 Arsenic plant tolerance improvement: increasing thiol-rich compound production

As(III) chelated with sulfhydryl-rich proteins forms complexes that get sequestered into the vacuoles through vacuolar transporters. Therefore, arsenic tolerance in plants can be enhanced by modifying GSH and PCs [97]. The constitutive expression of the PC biosynthetic gene *AtPCS1* under the CaMV 35S promoter in the non-accumulator plant *Nicotiana tabacum* enhanced PC levels. This resulted in an increased capacity of vacuolar arsenic and cadmium accumulation in roots and improved plant tolerance to these toxic elements [102]. The level of GSH, a precursor of PC, is a limiting factor for high PC production [103], so we suggest to combine *PCS1* overexpression along with the GSH biosynthetic genes *glutamate-cysteine ligase 1* (*GSH1*) or *glutathione synthetase 2* (*GSH2*) overexpression to maximize arsenic accumulation and tolerance.

In plants, arsenic exposure increases GSH content, which has been correlated with the feedback induction and increased expression of glutathione S-transferases

(GSTs) [104]. GSTs quench reactive molecules with the addition of GSH and protect the cell from oxidative damage. Indeed, overexpression of a rice *glutathione s-transferase* in *A. thaliana* led to increased tolerance toward arsenic and other abiotic stresses [105]. This may be attributed to the lower accumulation of reactive oxygen species and enhanced GST activity. Thus, gene expression of *GST* gene family could help to evolve strategies for developing arsenic-tolerant crops through biotechnological tools in the future.

4.1.3 Rhizoremediation: a beneficial plant-microbe interaction for arsenic phytoextraction

Rhizoremediation takes advantage of the interaction between plants and bacteria living in the rhizosphere for phytoremediation purposes [96, 106]. Arsenic-resistant bacteria associated to the rhizosphere have been demonstrated to play an important role in promoting plant growth and arsenic phytoextraction capacities from contaminated soils [107, 108]. Some bacteria and fungi increase the capability of plants to cope with arsenic by the release of phytohormones such as indole-3-acetic acid and/or essential vitamins and iron. These nutrients promote plant growth and reduce arsenic cytotoxicity. In addition, some microorganisms play an important role in arsenic bioavailability by catalyzing redox reactions that enhance the efficiency of arsenic uptake by the plant roots [109].

Bacterial strains isolated from the rhizosphere of autochthonous plants grown on arsenic-contaminated industrial lands have been used to enhance phytoremediation in *P. vittata* [108]. These arsenic-tolerant bacteria promoted plant growth and tolerance, as well as arsenic phytoextraction. This is possible due to plant growth elicitors released by microorganisms in particular siderophore-arsenic complexes, which are easily available for the plants promoting arsenic translocation to the shoots. Furthermore, arsenic can be volatilized by reducing As(V) and As(III) to the organic compound arsine, which can be easily taken up by the plant and released to the air through the leaves.

4.2 Development of arsenic-free food crops

Overall, the generation of arsenic-free crops in the aboveground organs involves increasing root sequestration or extrusion in order to reduce root-to-shoot translocation of the metalloid. Basic breakthroughs have been made in *A. thaliana* and other crop models, particularly in rice [25]. As already mentioned, rice naturally accumulates high amounts of arsenic in the form of As(III) when it is cultivated in paddy low-oxygen-containing soils, where As(III) is more prevalent [28]. For this reason, during the past decade, much research has been done to understand the mechanism of arsenic uptake, translocation, and grain filling. These studies are extremely useful for the production of rice with low concentrations of arsenic [25, 95].

4.2.1 Arsenate reductases and inositol transporters: from Arabidopsis to rice

AtARQ1/HAC1 constitutes the main As(V) reductase in the roots of *A. thaliana*. Knockout mutants of this gene show hypersensitivity to As(V) and increased translocation of arsenic to the shoots mediated by the phosphate transport system, leading to the hyperaccumulation of As(V) in aboveground tissues [55, 56]. Therefore, this As(V) reductase is essential to avoid root-to-shoot arsenic translocation. *OsHAC1;1*, *OsHAC1;2*, and *OsHAC4* are close rice homologues of *AtARQ1/HAC1*, highly expressed in roots. These genes are responsible for mitigating arsenic accumulation in rice shoots and grain. The overexpression of these rice As(V) reductases in roots results in grains exhibiting low concentrations of arsenic compared to wild-type grains [110, 111].

In *A. thaliana*, inositol transporters *AtINT2* and *AtINT4* are responsible for As(III) loading into the phloem and are key transporters regulating arsenic accumulation in plant seeds [112]. Thus, mutations in these transporters could potentially avoid arsenic accumulation in rice kernel. However, orthologues of inositol transporters in rice still need to be identified.

4.2.2 Arsenic extrusion strategies

In yeast, ACR3 from *S. cerevisiae* is the main As(III) efflux pump [48], and orthologues of this protein have not been found in *A. thaliana* or in rice [78]. In order to enhance arsenic efflux to the medium, yeast *ScACR3* was expressed in *A. thaliana*, leading to improved arsenic tolerance [113]. Importantly, *ScACR3* was also expressed in rice under the control of the CaMV 35S promoter, leading to a greater increase of As(III) efflux and mitigating arsenic accumulation in rice grain compared to wild-type plants [114]. Similarly, the heterologous expression in plants of the As(V) efflux transporter *ArsB* from *E. coli* could result in less arsenic accumulation. Hence, heterologous expression of well-characterized arsenic transporters from bacteria and fungi in plants could be a useful approach for crop improvement, since they allow the development of plants capable of growing in the presence of toxic levels of arsenic. Nevertheless, plants have a remarkable capacity to extrude As(III) from roots. Several transporters of the NIP aquaporin family are involved in a bidirectional transport of As(III) in plants [51]. We suggest to enhance As(III) efflux by coordinating an increased As(V) reduction through *AtARQ1* overexpression along with *NIP1;1* overexpression.

4.2.3 Decreasing arsenic accumulation in rice kernel: Lsi1 and Lsi2 knockout approach

In rice roots, the aquaporin OsNIP2;1/Lsi1 is a major entry route for Si(IV) [28]. This carrier also mediates As(III) uptake and methylated arsenic compounds, monomethylarsonic acid (MMA) and dimethylarsinic acid (DMA) [115]. Knockouts of this gene show significant lower levels of these arsenic species than wild-type rice plants at short times of exposure [116]. However at longer expositions, these knockout rice plants start to accumulate arsenic since there are bidirectional solute transporters driven by concentration gradient. Lsi2 is an additional aquaporin found in rice. Lsi2 mediates Si(IV) efflux and As(III) transport into the symplast avoiding the Casparian strip and then being easily loaded into the xylem [117]. A knockout mutation of this gene in rice leads to a dramatic reduction on As(III) transport to the xylem and its accumulation in the shoots and grains [117].

4.2.4 Arsenic sequestration in root vacuoles to avoid translocation

An *ACR3* orthologue that mediates As(III) transport into the vacuole was recently identified in *P. vittata* [78]. When ectopically overexpressed in *A. thaliana*, it significantly favors arsenic translocation to the aerial parts instead of the accumulation in the root vacuoles [118]. Therefore, it seems to be a good option to highly express this vacuolar transporter specifically in the root tissues in order to sequester arsenic in the root vacuoles and prevent As(III) from being translocated to the shoot and grains [97]. Additionally, the rice vacuolar transporter OsABCC1, an *A. thaliana* ABCC1 orthologue, is involved in As(III)-PC sequestration inside the vacuoles as well; thus the overexpression of this transporter in roots could help to limit arsenic content in the grains [119]. Moreover, it has been recently shown that transgenic rice overexpressing *OsPCS1* showed significantly lower arsenic content in grains [120]. These results provide a promising strategy to decrease arsenic translocation by making use of vascular transporters and PC overproducers to obtain low-arsenic crops.

4.2.5 Conversion of arsenic to less toxic volatile forms

Microorganisms are able to volatilize metals and metalloids into the atmosphere through methylation [40]. Several phytoremediation approaches for mercury and arsenic volatilization have been proposed as a suitable strategy to produce safe crops grown in contaminated crops. However, plants do not have in their genome genes encoding As(III) methyltransferases [25]. Therefore, the methylated species of arsenic (MMAs and DMAs) found in plants are the result of microbial activity in the rhizosphere. The volatile arsenic species produced by these microorganisms are then taken up, translocated, and released to the atmosphere by the plant [25]. The reduction of arsenic accumulation through volatilization in rice has been achieved by overexpression of As(III) methyltransferases from bacteria and fungi [121, 122]. Although this approach is very efficient, the release of contaminants into the atmosphere is controversial.

4.2.6 Controlling the AsV/Pi transporter specificity

It has been recently shown that the phosphate transporter PvPHT1;2 from *P. vittata* displays a significantly lower As(V) uptake than other phosphate transporters [123]. The heterologous expression of this transporter in tobacco plants resulted in plants with increased shoot biomass in the presence of As(V) due to lower arsenic accumulation. Thus, it is tempting to speculate that genome-wide analysis of phosphate transporters from several species with different accumulation rates will allow the identification of new alleles that favor As(V) or phosphate uptake for phytoremediation or food safety purposes, respectively. Importantly, discovering the mechanisms that modulate substrate specificity of the arsenic transporters would be essential to develop safer crops or hyperaccumulators, which would specifically release or accumulate arsenic, respectively, without interfering with the homeostasis of essential metals or phosphate [97].

4.2.7 Phytochelatin and GSH pathway to decrease As accumulation

As mentioned above, one of the most important peptides involved in arsenic accumulation and detoxification of arsenic is the PC, forming As(III)-PC complexes that can be sequestered in the vacuoles. Endosperm-specific intron-containing hairpin RNA-mediated gene silencing of *OsPCS1* and *OsPCS2* resulted in lower arsenic sequestration in the seed vacuoles by PCs in the transgenic rice [124]. This led to the accumulation of significantly less arsenic in the grains than those in the wild type. GRXs are cysteine-rich proteins and GSH-dependent redox enzymes in charge of protection against oxidative stress, maintaining the intracellular GSH pool [125]. Recently, two arsenic-responsive rice GRX families of proteins from rice, OsGrx_C7 and OsGrx_C2.1, have been reported [126]. The heterologous overexpression of these proteins in *A. thaliana* increased plant tolerance to arsenic and reduced As(III) shoot and seed accumulation. This low As(III) accumulation in rice can be due to the fact that GRXs regulate intracellular As(III) levels by an indirect or direct modulation of the NIP transporters involved in As(III) uptake and extrusion [126]. Understanding arsenic-induced stress pathways is of central importance for the design of rice crops displaying less arsenic content in their edible parts either through the upregulation of *OsGrxs* or the seed-specific downregulation of *OsPCS*.

4.2.8 Iron plaques on rice roots restrict arsenic availability

The oxygenation of roots as a consequence of oxygen diffusion from aerenchyma to the rhizosphere and microbial growth leads to the oxidation of iron and the subsequent formation of iron plaques on the surface of roots particularly in rice [127]. The formation of these plaques is highly variable among different rice genotypes [128]. These iron plaques promote arsenic adsorption and sequestration on the root surface through the formation of Fe-As complexes, thus restricting arsenic bioavailability for plants. Hence, there is a direct relationship between iron plaque formation and reduced arsenic accumulation in the aboveground biomass in rice [128]. Elucidating the genetic mechanisms that affect iron plaque formation would encourage the generation of genetically modified rice with low arsenic accumulation.

4.2.9 Transcription factors: the key coordinators of the arsenic response

Transcriptional regulation is a major factor in the regulation of the capacity of plants to tolerate and accumulate arsenic [99]. Therefore a suitable approach to enhance plant phytoremediation performance will be the identification of key transcriptional regulators of the arsenic response. In *A. thaliana* two WRKY transcription factors have been identified as regulators of As(V) uptake. On the one hand, WRKY45 activates *PHT1;1* expression in response to phosphate starvation, mediating the As(V) influx as a phosphate analogue [129]. On the other hand, WRKY6, an As(V)-responsive transcription factor restricts As(V) uptake by the downregulation of *PHT1;1* [130]. Recently, *OsARM1*, a MYB transcription factor, has been identified in rice. This gene is strongly induced by As(III) and negatively regulates arsenic-associated transporters genes, namely, *OsLsi1*, *OsLsi2*, and *OsLsi6*, having a potential role in the transcriptional regulation of arsenic response in rice [131].

A coordinated network of arsenic, transport, chelation, trafficking, and sequestration mechanisms is crucial to uptake, translocate, and detoxify arsenic. To achieve this, there must be a strong transcriptional regulation following arsenic exposure [99]. However, sensing of arsenic and the signal transduction pathway remains completely unknown. As a matter of fact, a master regulator controlling the expression of other key transcription factor molecules in response to arsenic has not been identified yet. Discovering such transcription factor would be fundamental for developing genetically modified crops that trigger the expression of arsenic detoxification genes to adapt plants to the stress in a synchronized manner.

4.2.10 Relevance of transcriptomic approaches for phytoremediation strategies

The basic discoveries of the arsenic signaling pathway drawn from model plants such as *A. thaliana* hold great potential for the development of new phytoremediation strategies using crop plants. Recent studies are taking advantage of "omic" technologies to elucidate the genetic regulators and pathways responsible for arsenic response in plants. Cadmium and arsenic are toxic elements in rice that often appear together in contaminated paddy field soils [25]. To address whether rice has a common molecular response mechanism against both cadmium and arsenic toxicity, the identification of key genes—expressed in response to these contaminant elements—was recently aimed through a transcriptomic analysis by RNA-sequencing [132]. They found that the vast majority of the genes that responded to both arsenic and cadmium fall into five categories: redox-, glutathione metabolism-, cell wall biogenesis-, expression regulation-, and transmembrane transport-related genes. From these, they selected the most differentially expressed genes and confirmed 27 common responsive genes to both As and Cd stress by RT-qPCR [132]. Additionally, our group performed an RNA-seq of *A. thaliana* seedlings exposed to As(V) (unpublished results), in order to provide a deeper understanding of the molecular mechanisms responsible for the arsenic stress response. **Table 1** shows the 27 arsenic- and cadmium-responsive genes in rice and their corresponding homologous arsenic-responsive genes in *A. thaliana*. The conclusion to be drawn from this is that most of the arsenic- and cadmium-responsive genes found in both *A. thaliana* and rice belong to redox and glutathione metabolism, as well as transmembrane transport, suggesting a conservation of these processes between these species; thus, carrying out research in both species is extremely useful for crop development and improvement.

	Arsenic- and cadmium-responsive gene in rice	**Arsenic-responsive orthologue in *A. thaliana***
Redox-related genes	Os06g0216000	
	Os07g0638300	
	Os07g0418500	AT2G46950
	Os03g0227700	AT3G50660
	Os01g0294500	AT2G38380
Glutathione metabolism-related genes	Os09g0367700	AT2G29490, AT2G29450
	Os03g0283200	AT5G02780
	Os10g0530900	AT2G29440
	Os12g0263000	AT5G27380
Cell wall biogenesis-related genes	Os11g0592000	
	Os05g0247800	
	Os03g0155700	AT2G03090
	Os03g0416200	AT5G15630
Expression regulation-related genes	Os02g0168200	AT1G14600
	Os07g0129200	
	Os07g0597200	AT1G74360
	Os08g0203300	
	Os09g0423200	
	Os02g0557800	AT1G59940
	Os06g0692500	
Transmembrane transport-related genes	Os01g0695800	AT2G36910
	Os04g0524500	AT1G65730
	Os01g0939100	AT2G41560
	Os01g0307500	AT4G10310
	Os01g0247700	AT1G62262
	Os01g0623200	AT4G27970
	Os03g0107300	

Table 1.
Cadmium-responsive genes in rice and their corresponding homologous arsenic-responsive genes in A. thaliana.

5. Conclusions

Arsenic contamination poses a global threat for all living organisms; for this reason, different strategies have been developed to cope with this serious challenge. Among them, phytoremediation is a promising approach. In this chapter, we have provided a brief overview on the status of arsenic phytoremediation from polluted soils and waters, with special focus on the mechanisms of arsenic perception and tolerance in several organisms, plant species used for arsenic phytoremediation, and biotechnological approaches that are driven to increase phytoremediation efficiency as well as crop protection. Although much knowledge and experience have been gained over the last years, there are still many aspects to be discovered and improved. We have exciting times ahead: the exploitation of natural variation, the use of "omic" technologies and biotechnological approaches, a holistic perspective of plant-soil-microbiota interactions, and valorization of plant-fixed arsenic will provide unique opportunities to boost this green strategy.

Acknowledgements

We apologize to all colleagues whose studies were not cited in this review because of space constraints. We gratefully acknowledge funding by the Spanish Ministry of Economy and Competitiveness (project BIO2014-55741-R) and by the LIFE Programme of the European Commission (Grant number LIFE15 ENV/ES/000382) to AL. CN was supported by project BIO2014-55741-R, CM was supported by project LIFE15 ENV/ES/000382, and MN was supported by a FPU grant (FPU17/05202) from the Spanish Ministry of Science and Innovation.

Conflict of interest

The authors declare no conflict of interest.

Author details

Cristian Mateo†, Micaela Navarro†, Cristina Navarro† and Antonio Leyva*
Department of Plant Molecular Genetics, National Centre of Biotechnology, Spanish National Research Council (CSIC), Madrid, Spain

*Address all correspondence to: aleyva@cnb.csic.es

†These authors are contributed equally to this work.

References

[1] Mandal BK, Suzuki KT. Arsenic round the world: A review. Talanta. 2002;**58**:201-235. DOI: 10.1016/S0039-9140(02)00268-0

[2] Chen SJ, Yan XJ, Chen Z. Arsenic in nature. In: Encyclopedia of Metalloproteins. New York, NY: Springer; 2013. pp. 120-123. DOI: 10.1007/978-1-4614-1533-6_489

[3] Su C, Jiang L, Zhang W. A review on heavy metal contamination in the soil worldwide: Situation, impact and remediation techniques. Environmental Skeptics and Critics. 2014;**3**:24-38. DOI: 10.1037/a0036071

[4] Shen S, Li X-F, Cullen WR, Weinfeld M, Le XC. Arsenic binding to proteins. Chemical Reviews. 2013;**113**:7769-7792. DOI: 10.1021/cr300015c

[5] Tchounwou PB, Yedjou CG, Udensi UK, Pacurari M, Stevens JJ, Patlolla AK, et al. State of the science review of the health effects of inorganic arsenic: Perspectives for future research. Environmental Toxicology. 2019;**34**:188-202. DOI: 10.1002/tox.22673

[6] Ratnaike RN. Acute and chronic arsenic toxicity. Postgraduate Medical Journal. 2003;**79**:391-396. DOI: 10.1136/pmj.79.933.391

[7] Jomova K, Jenisova Z, Feszterova M, Baros S, Liska J, Hudecova D, et al. Arsenic: Toxicity, oxidative stress and human disease. Journal of Applied Toxicology. 2011;**31**:95-107. DOI: 10.1002/jat.1649

[8] Mohammed Abdul KS, Jayasinghe SS, Chandana EPS, Jayasumana C, De Silva PMCS. Arsenic and human health effects: A review. Environmental Toxicology and Pharmacology. 2015;**40**(3):828-846. DOI: 10.1016/j.etap.2015.09.016

[9] Peidang L, Kewen H, Yinxia L, Qiuli W, Peng Y, Dayong W. Exposure to mercury causes formation of male-specific structural deficits by inducing oxidative damage in nematodes. Ecotoxicology and Environmental Safety. 2012;**79**:90-100. DOI: 10.1016/j.ecoenv.2011.12.007

[10] Hong YS, Song KH, Chung JY. Health effects of chronic arsenic exposure. Journal of Preventive Medicine & Public Health. 2014;**47**:245-252. DOI: 10.3961/jpmph.14.035

[11] Smith AH, Lingas EO, Rahman M. Contamination of drinking-water by arsenic in Bangladesh: A public health emergency. Bulletin of the World Health Organization. 2000;**78**:1093-1103

[12] Chakraborti D, Rahman MM, Mukherjee A, Alauddin M, Hassan M, Dutta RN, et al. Groundwater arsenic contamination in Bangladesh-21 years of research. Journal of Trace Elements in Medicine and Biology. 2015;**31**:237-248. DOI: 10.1016/j.jtemb.2015.01.003

[13] Banerjee M, Banerjee N, Bhattacharjee P, Mondal D, Lythgoe PR, Martínez M, et al. High arsenic in rice is associated with elevated genotoxic effects in humans. Scientific Reports. 2013;**3**:2195. DOI: 10.1038/srep02195

[14] Hojsak I, Braegger C, Bronsky J, Campoy C, Colomb V, Decsi T, et al. Arsenic in rice: A cause for concern. Journal of Pediatric Gastroenterology and Nutrition. 2015;**60**(1):142-145. DOI: 10.1097/MPG.0000000000000502

[15] Gilbert-Diamond D, Cottingham KL, Gruber JF, Punshon T, Sayarath V, Gandolfi AJ, et al. Rice consumption contributes to arsenic exposure in US women. Proceedings of the National Academy of Sciences. 2011;**108**:20656-20660. DOI: 10.1073/pnas.1109127108

[16] Cascio C, Raab A, Jenkins RO, Feldmann J, Meharg AA, Haris PI. The impact of a rice based diet on urinary arsenic. Journal of Environmental Monitoring. 2011;**13**:257-265. DOI: 10.1039/c0em00482k

[17] Mitra A, Chatterjee S, Moogouei R, Gupta D. Arsenic accumulation in rice and probable mitigation approaches: A review. Agronomy. 2017;7(4):67. DOI: 10.3390/agronomy7040067

[18] Sarkar A, Paul B. The global menace of arsenic and its conventional remediation—A critical review. Chemosphere. 2016;**158**:37-49. DOI: 10.1016/j.chemosphere.2016.05.043

[19] Nicomel NR, Leus K, Folens K, Van Der Voort P, Du Laing G. Technologies for arsenic removal from water: Current status and future perspectives. International Journal of Environmental Research and Public Health. 2016;**13**:62. DOI: 10.3390/ijerph13010062

[20] Hayat K, Menhas S, Bundschuh J, Chaudhary HJ. Microbial biotechnology as an emerging industrial wastewater treatment process for arsenic mitigation: a critical review. Journal of Cleaner Production. 2017;**151**:427-438

[21] Surriya O, Sarah Saleem S, Waqar K, Gul Kazi A. Phytoremediation of soils. In: Soil Remediation and Plants: Prospects and Challenges. Amsterdam: Elsevier; 2014. DOI: 10.1016/B978-0-12-799937-1.00001-2

[22] Sarwar N, Imran M, Shaheen MR, Ishaque W, Kamran MA, Matloob A, et al. Phytoremediation strategies for soils contaminated with heavy metals: Modifications and future perspectives. Chemosphere. 2017;**171**:710-721. DOI: 10.1016/j.chemosphere.2016.12.116

[23] Lee JH. An overview of phytoremediation as a potentially promising technology for environmental pollution control. Biotechnology and Bioprocess Engineering. 2013;**18**:431-439. DOI: 10.1007/s12257-013-0193-8

[24] Eze VC, Harvey AP. Extractive recovery and valorisation of arsenic from contaminated soil through phytoremediation using *Pteris cretica*. Chemosphere. 2018;**208**:484-492. DOI: 10.1016/j.chemosphere.2018.06.027

[25] Clemens S, Ma JF. Toxic heavy metal and metalloid accumulation in crop plants and foods. Annual Review of Plant Biology. 2016;**67**:489-512. DOI: 10.1146/annurev-arplant-043015-112301

[26] Navarro C, Arbaoui S, Mateo C, Bettaieb T, Leyva A. Arsenic, the Silent Threat: New Phytoremediation Strategies for Contaminated Soils and Waters. Hauppauge: Nova Science Pub Inc.; 2018

[27] Mendoza-Cózatl DG, Jobe TO, Hauser F, Schroeder JI. Long-distance transport, vacuolar sequestration, tolerance, and transcriptional responses induced by cadmium and arsenic. Current Opinion in Plant Biology. 2011;**14**(5):554-562. DOI: 10.1016/j.pbi.2011.07.004

[28] Zhao FJ, McGrath SP, Meharg AA. Arsenic as a food chain contaminant: Mechanisms of plant uptake and metabolism and mitigation strategies. Annual Review of Plant Biology. 2010;**61**:535-559. DOI: 10.1146/annurev-arplant-042809-112152

[29] Novick RP, Roth C. Plasmid-linked resistance to inorganic salts in *Staphylococcus aureus*. Journal of Bacteriology. 1968;**95**:1335-1342

[30] Hedges RW, Baumberg S. Resistance to arsenic compounds conferred by a plasmid transmissible between strains of *Escherichia coli*. Journal of Bacteriology. 1973;**115**:459-460

[31] San Francisco MJD, Hope CL, Owolabi JB, Tisa LS, Rosen BP. Identification of the metalloregulatory element of the plasmid-encoded arsenical resistance operon. Nucleic Acids Research. 1990;**18**:619-624. DOI: 10.1093/nar/18.3.619

[32] Ben FI, Zhang C, Li YP, Zhao Y, Alwathnani HA, Saquib Q, et al. Distribution of arsenic resistance genes in prokaryotes. Frontiers in Microbiology. 2018;**9**:2473. DOI: 10.3389/fmicb.2018.02473

[33] Wu J, Rosen BP. The ArsR protein is a trans-acting regulatory protein. Molecular Microbiology. 1991;**5**: 1331-1336. DOI: 10.1111/j.1365-2958.1991.tb00779.x

[34] Busenlehner LS, Pennella MA, Giedroc DP. The SmtB/ArsR family of metalloregulatory transcriptional repressors: Structural insights into prokaryotic metal resistance. FEMS Microbiology Reviews. 2003;**27**:131-143. DOI: 10.1016/S0168-6445(03)00054-8

[35] Shi W, Dong J, Scott RA, Ksenzenko MY, Rosen BP. The role of arsenic-thiol interactions in metalloregulation of the ars operon. The Journal of Biological Chemistry. 1996;**271**:9291-9297. DOI: 10.1074/jbc.271.16.9291

[36] Xu C, Rosen BP. Dimerization is essential for DNA binding and repression of the trans-acting metalloregulatory ArsR protein of *Escherichia coli*. The FASEB Journal. 1997;**11**:15734-15738. DOI: 10.1074/jbc.272.25.15734

[37] Ordóñez E, Thiyagarajan S, Cook JD, Stemmler TL, Gil JA, Mateos LM, et al. Evolution of metal(loid) binding sites in transcriptional regulators. The Journal of Biological Chemistry. 2008;**283**:25706-25714. DOI: 10.1074/jbc.M803209200

[38] Antonucci I, Gallo G, Limauro D, Contursi P, Ribeiro AL, Blesa A, et al. An ArsR/SmtB family member regulates arsenic resistance genes unusually arranged in *Thermus thermophilus* HB27. Microbial Biotechnology. 2017;**10**:1690-1701. DOI: 10.1111/1751-7915.12761

[39] Gladysheva TB, Oden KL, Rosen BP. Properties of the arsenate reductase of plasmid R773. Biochemistry. 1994;**33**:7288-7293. DOI: 10.1021/bi00189a033

[40] Zhu Y-G, Yoshinaga M, Zhao F-J, Rosen BP. Earth abides arsenic biotransformations. Annual Review of Earth and Planetary Sciences. 2014;**42**:443-467. DOI: 10.1146/annurev-earth-060313-054942

[41] Meng YL, Liu Z, Rosen BP. As(III) and Sb(III) uptake by GlpF and efflux by ArsB in *Escherichia coli*. The Journal of Biological Chemistry. 2004;**279**:18334-18341. DOI: 10.1074/jbc.M400037200

[42] Chen Y, Rosen BP. Metalloregulatory properties of the ArsD repressor. The Journal of Biological Chemistry. 1997;**272**:14257-14262. DOI: 10.1074/jbc.272.22.14257

[43] Lin Y-F, Walmsley AR, Rosen BP. An arsenic metallochaperone for an arsenic detoxification pump. Proceedings of the National Academy of Sciences. 2006;**103**:15617-15622. DOI: 10.1073/pnas.0603974103

[44] Yang J, Rawat S, Stemmler TL, Rosen BP. Arsenic binding and transfer by the ArsD as (III) metallochaperone. Biochemistry. 2010;**49**:3658-3666. DOI: 10.1021/bi100026a

[45] West KL, Byrum SD, Mackintosh SG, Edmondson RD, Taverna SD, Tackett AJ. Proteomic characterization of the arsenic response locus in *S. cerevisiae*. Epigenetics. 2019;**14**:130-145. DOI: 10.1080/15592294.2019.1580110

[46] Kumar NV, Yang J, Pillai JK, Rawat S, Solano C, Kumar A, et al. Arsenic directly binds to and activates the yeast AP-1-like transcription factor Yap8. Molecular and Cellular Biology. 2015;**36**:913-922. DOI: 10.1128/mcb.00842-15

[47] Mukhopadhyay R, Rosen BP. *Saccharomyces cerevisiae* ACR2 gene encodes an arsenate reductase. FEMS Microbiology Letters. 1998;**168**:127-136. DOI: 10.1016/S0378-1097(98)00430-3

[48] Wysocki R, Bobrowicz P, Ułaszewski S. The *Saccharomyces cerevisiae* ACR3 gene encodes a putative membrane protein involved in arsenite transport. The Journal of Biological Chemistry. 1997;**272**:30061-30066. DOI: 10.1074/jbc.272.48.30061

[49] Catarecha P, Segura MD, Franco-Zorrilla JM, Garcia-Ponce B, Lanza M, Solano R, et al. A mutant of the *Arabidopsis* phosphate transporter PHT1;1 displays enhanced arsenic accumulation. The Plant Cell. 2007;**19**:1123-1133. DOI: 10.1105/tpc.106.041871

[50] Kamiya T, Tanaka M, Mitani N, Jian FM, Maeshima M, Fujiwara T. NIP1;1, an aquaporin homolog, determines the arsenite sensitivity of *Arabidopsis thaliana*. The Journal of Biological Chemistry. 2009;**284**:2114-2120. DOI: 10.1074/jbc.M806881200

[51] Pommerrenig B, Diehn TA, Bienert GP. Metalloido-porins: Essentiality of nodulin 26-like intrinsic proteins in metalloid transport. Plant Science. 2015;**238**:212-227. DOI: 10.1016/j.plantsci.2015.06.002

[52] Xu W, Dai W, Yan H, Li S, Shen H, Chen Y, et al. *Arabidopsis* NIP3;1 plays an important role in arsenic uptake and root-to-shoot translocation under arsenite stress conditions. Molecular Plant. 2015;**8**:722-733. DOI: 10.1016/j.molp.2015.01.005

[53] Dhankher OP, Rosen BP, McKinney EC, Meagher RB. Hyperaccumulation of arsenic in the shoots of *Arabidopsis* silenced for arsenate reductase (ACR2). Proceedings of the National Academy of Sciences. 2006;**103**:5413-5418. DOI: 10.1073/pnas.0509770102

[54] Liu W, Schat H, Bliek M, Chen Y, McGrath SP, George G, et al. Knocking out ACR2 does not affect arsenic redox status in *Arabidopsis thaliana*: Implications for as detoxification and accumulation in plants. PLoS One. 2012;**7**(8):e42408. DOI: 10.1371/journal.pone.0042408

[55] Sánchez-Bermejo E, Castrillo G, Del Llano B, Navarro C, Zarco-Fernández S, Martinez-Herrera DJ, et al. Natural variation in arsenate tolerance identifies an arsenate reductase in *Arabidopsis thaliana*. Nature Communications. 2014;**5**:4617. DOI: 10.1038/ncomms5617

[56] Chao DY, Chen Y, Chen J, Shi S, Chen Z, Wang C, et al. Genome-wide association mapping identifies a new arsenate reductase enzyme critical for limiting arsenic accumulation in plants. PLoS Biology. 2014;**12**:e1002009. DOI: 10.1371/journal.pbio.1002009

[57] Wood BA, Feldmann J. Quantification of phytochelatins and their metal(loid) complexes: Critical assessment of current analytical methodology. Analytical and Bioanalytical Chemistry. 2012;**402**:3299-3309. DOI: 10.1007/s00216-011-5649-0

[58] Sharma SS, Dietz KJ, Mimura T. Vacuolar compartmentalization as indispensable component of heavy metal detoxification in plants. Plant Cell & Environment. 2016;**39**:1112-1126. DOI: 10.1111/pce.12706

[59] Song W-Y, Park J, Mendoza-Cozatl DG, Suter-Grotemeyer M, Shim D, Hortensteiner S, et al. Arsenic tolerance in *Arabidopsis* is mediated by two ABCC-type phytochelatin transporters.

Proceedings of the National Academy of Sciences. 2010;**107**:21187-21192. DOI: 10.1073/pnas.1013964107

[60] Luan M, Liu J, Liu Y, Han X, Sun G, Lan W, et al. Vacuolar phosphate transporter 1 (VPT1) affects arsenate tolerance by regulating phosphate homeostasis in *Arabidopsis*. Plant & Cell Physiology. 2018;**59**:1345-1352. DOI: 10.1093/pcp/pcy025

[61] Reeves RD, Baker AJM, Jaffré T, Erskine PD, Echevarria G, Van der Ent A. A global database for plants that hyperaccumulate metal and metalloid trace elements. The New Phytologist. 2018;**218**:407-411. DOI: 10.1111/nph.14907

[62] Van der Ent A, Baker AJM, Reeves RD, Pollard AJ, Schat H. Hyperaccumulators of metal and metalloid trace elements: Facts and fiction. Plant and Soil. 2013;**362**: 319-334. DOI: 10.1007/s11104-012-1287-3

[63] Niazi NK, Bashir S, Bibi I, Murtaza B, Shahid M, Javed MT. Phytoremediation of arsenic-contaminated soils using arsenic hyperaccumulating ferns. In: Ansari A, Gill S, Gill R, Lanza G, Newman L, editors. Phytoremediation. Cham: Springer; 2016. pp. 521-545. DOI: 10.1007/978-3-319-40148-5_19

[64] Rathinasabapathi B, Ma LQ, Srivastava M. Arsenic hyper-accumulating ferns and their application to phytoremediation of arsenic contaminated sites. In: Teixeira da Silva JA, editor. Floricultural Advances. Ikenobe, Japan: Global Science Books; 2006. pp. 304-311

[65] da Silva EB, Lessl JT, Wilkie AC, Liu X, Liu Y, Ma LQ. Arsenic removal by As-hyperaccumulator *Pteris vittata* from two contaminated soils: A 5-year study. Chemosphere. 2018;**206**:736-741. DOI: 10.1016/j.chemosphere.2018.05.055

[66] Ye WL, Khan MA, McGrath SP, Zhao FJ. Phytoremediation of arsenic contaminated paddy soils with *Pteris vittata* markedly reduces arsenic uptake by rice. Environmental Pollution. 2011;**159**:3739-3743. DOI: 10.1016/j.envpol.2011.07.024

[67] Wan XM, Lei M, Liu YR, Huang ZC, Chen TB, Gao D. A comparison of arsenic accumulation and tolerance among four populations of *Pteris vittata* from habitats with a gradient of arsenic concentration. Science of the Total Environment. 2013;**442**:143-151. DOI: 10.1016/j.scitotenv.2012.10.056

[68] Natarajan S, Stamps RH, Ma LQ, Saha UK, Hernandez D, Cai Y, et al. Phytoremediation of arsenic-contaminated groundwater using arsenic hyperaccumulator *Pteris vittata* L.: Effects of frond harvesting regimes and arsenic levels in refill water. Journal of Hazardous Materials. 2011;**185**:983-989. DOI: 10.1016/j.jhazmat.2010.10.002

[69] Bondada BR, Ma LQ. Tolerance of heavy metals in vascular plants: Arsenic hyperaccumulation by Chinese brake fern (*Pteris Vittata* L.). In: Chandra S, Srivastava M, editors. Pteridology in the New Millennium. Dordrecht: Springer; 2003. pp. 397-420. DOI: 10.1007/978-94-017-2811-9_28

[70] Ma LQ, Komar KM, Tu C, Zhang W, Cai Y, Kennelley ED. A fern that hyperaccumulates arsenic. Nature. 2001;**409**:579-579. DOI: 10.1038/35054664

[71] Ma J, Lei E, Lei M, Liu Y, Chen T. Remediation of arsenic contaminated soil using malposed intercropping of *Pteris vittata* L. and maize. Chemosphere. 2018;**194**:737-744. DOI: 10.1016/j.chemosphere.2017.11.135

[72] Wan X, Lei M, Chen T, Yang J. Intercropped *Pteris vittata* L. and *Morus alba* L. presents a safe utilization mode for arsenic-contaminated soil.

Science of the Total Environment. 2017;**579**:1467-1475. DOI: 10.1016/j.scitotenv.2016.11.148

[73] Wan X, Lei M. Intercropping efficiency of four arsenic hyperaccumulator *Pteris vittata* populations as intercrops with *Morus alba*. Environmental Science and Pollution Research International. 2018;**25**:12600-12611. DOI: 10.1007/s11356-018-1366-y

[74] Ditusa SF, Fontenot EB, Wallace RW, Silvers MA, Steele TN, Elnagar AH, et al. A member of the phosphate transporter 1 (Pht1) family from the arsenic-hyperaccumulating fern *Pteris vittata* is a high-affinity arsenate transporter. The New Phytologist. 2016;**209**:762-772. DOI: 10.1111/nph.13472

[75] He Z, Yan H, Chen Y, Shen H, Xu W, Zhang H, et al. An aquaporin PvTIP4;1 from *Pteris vittata* may mediate arsenite uptake. The New Phytologist. 2016;**209**:746-761. DOI: 10.1111/nph.13637

[76] Cesaro P, Cattaneo C, Bona E, Berta G, Cavaletto M. The arsenic hyperaccumulating *Pteris vittata* expresses two arsenate reductases. Scientific Reports. 2015;**5**:14525. DOI: 10.1038/srep14525

[77] Rathinasabapathi B, Wu S, Sundaram S, Rivoal J, Srivastava M, Ma LQ. Arsenic resistance in *Pteris vittata* L.: Identification of a cytosolic triosephosphate isomerase based on cDNA expression cloning in *Escherichia coli*. Plant Molecular Biology. 2006;**62**:845-857. DOI: 10.1007/s11103-006-9060-8

[78] Indriolo E, Na G, Ellis D, Salt DE, Banks JA. A vacuolar arsenite transporter necessary for arsenic tolerance in the arsenic hyperaccumulating fern *Pteris vittata* is missing in flowering plants. The Plant Cell. 2010;**22**:2045-2057. DOI: 10.1105/tpc.109.069773

[79] Shen H, He Z, Yan H, Xing Z, Chen Y, Xu W, et al. The fronds tonoplast quantitative proteomic analysis in arsenic hyperaccumulator *Pteris vittata* L. Journal of Proteomics. 2014;**105**:46-57. DOI: 10.1016/j.jprot.2014.01.029

[80] Srivastava M, Ma LQ, Singh N, Singh S. Antioxidant responses of hyper-accumulator and sensitive fern species to arsenic. Journal of Experimental Botany. 2005;**56**:1335-1342. DOI: 10.1093/jxb/eri134

[81] Potdukhe RM, Bedi P, Sarangi BK, Pandey RA, Thul ST. Root transcripts associated with arsenic accumulation in hyperaccumulator *Pteris vittata*. Journal of Biosciences. 2018;**43**:105-115. DOI: 10.1007/s12038-018-9735-8

[82] Yan H, Gao Y, Wu L, Wang L, Zhang T, Dai C, et al. Potential use of the *Pteris vittata* arsenic hyperaccumulation-regulation network for phytoremediation. Journal of Hazardous Materials. 2019;**368**:386-396. DOI: 10.1016/j.jhazmat.2019.01.072

[83] Liu X, Feng HY, Fu JW, Sun D, Cao Y, Chen Y, et al. Phytate promoted arsenic uptake and growth in arsenic-hyperaccumulator *Pteris vittata* by upregulating phosphorus transporters. Environmental Pollution. 2018;**241**:240-246. DOI: 10.1016/j.envpol.2018.05.054

[84] Chu H, Vu T, Nguyen T, Nguyen H. Accumulation of arsenic and heavy metals in native and cultivated plant species in a lead recycling area in Vietnam. Minerals. 2019;**9**:132. DOI: 10.3390/min9020132

[85] Mahmud R, Inoue N, Kasajima SY, Shaheen R. Assessment of potential indigenous plant species for the phytoremediation of arsenic-contaminated areas of Bangladesh. International Journal of

Phytoremediation. 2008;**10**:117-130. DOI: 10.1080/15226510801913884

[86] Anh BTK, Kim DD, Van Tua T, Kien NT, Anh DT. Phytoremediation potential of indigenous plants from Thai Nguyen province, Vietnam. Journal of Environmental Biology. 2011;**32**:257-262

[87] Visoottiviseth P, Francesconi K, Sridokchan W. The potential of Thai indigenous plant species for the phytoremediation of arsenic contaminated land. Environmental Pollution. 2002;**118**:453-461. DOI: 10.1016/S0269-7491(01)00293-7

[88] Robinson B, Kim N, Marchetti M, Moni C, Schroeter L, van den Dijssel C, et al. Arsenic hyperaccumulation by aquatic macrophytes in the Taupo Volcanic Zone, New Zealand. Environmental and Experimental Botany. 2006;**58**:206-215. DOI: 10.1016/j.envexpbot.2005.08.004

[89] Robinson B, Marchetti M, Moni C, Schroeter L, van den Dijssel C, Milne G, et al. Arsenic accumulation by aquatic and terrestrial plants. In: Managing Arsenic in the Environment: From Soil to Human Health. Collingwood, Victoria: CSIRO; 2006. pp. 235-247

[90] Jasrotia S, Kansal A, Mehra A. Performance of aquatic plant species for phytoremediation of arsenic-contaminated water. Applied Water Science. 2015;**7**:889-896. DOI: 10.1007/s13201-015-0300-4

[91] Li B, Gu B, Yang Z, Zhang T. The role of submerged macrophytes in phytoremediation of arsenic from contaminated water: A case study on *Vallisneria natans* (Lour.) Hara. Ecotoxicology and Environmental Safety. 2018;**165**:224-231. DOI: 10.1016/j.ecoenv.2018.09.023

[92] Rahman MA, Hasegawa H. Aquatic arsenic: Phytoremediation using floating macrophytes. Chemosphere. 2011;**83**:63346. DOI: 10.1016/j.chemosphere.2011.02.045

[93] Jomjun N, Siripen T, Maliwan S, Jintapat N, Prasak T, Somporn C, et al. Phytoremediation of arsenic in submerged soil by wetland plants. International Journal of Phytoremediation. 2011;**13**:35-46. DOI: 10.1080/15226511003671320

[94] Mkandawire M, Taubert B, Dudel EG. Capacity of *Lemna gibba* L. (Duckweed) for uranium and arsenic phytoremediation in mine tailing waters. International Journal of Phytoremediation. 2004;**6**:347-362. DOI: 10.1080/16226510490888884

[95] Shri M, Singh PK, Kidwai M, Gautam N, Dubey S, Verma G, et al. Recent advances in arsenic metabolism in plants: Current status, challenges and highlighted biotechnological intervention to reduce grain arsenic in rice. Metallomics. 2019;**11**:519-532. DOI: 10.1039/c8mt00320c

[96] Mosa KA, Saadoun I, Kumar K, Helmy M, Dhankher OP. Potential biotechnological strategies for the cleanup of heavy metals and metalloids. Frontiers in Plant Science. 2016;7:303. DOI: 10.3389/fpls.2016.00303

[97] Zhang J, Martinoia E, Lee Y. Vacuolar transporters for cadmium and arsenic in plants and their applications in phytoremediation and crop development. Plant & Cell Physiology. 2018;**59**:1317-1325. DOI: 10.1093/pcp/pcy006

[98] Garbinski LD, Rosen BP, Chen J. Pathways of arsenic uptake and efflux. Environment International. 2019;**126**:585-597. DOI: 10.1016/j.envint.2019.02.058

[99] Clemens S. Molecular mechanisms of plant metal tolerance and homeostasis. Planta. 2001;**212**:475-486. DOI: 10.1007/s004250000458

[100] Wang C, Na G, Bermejo ES, Chen Y, Banks JA, Salt DE, et al. Dissecting the components controlling root-to-shoot arsenic translocation in *Arabidopsis thaliana*. The New Phytologist. 2018;**217**:206. DOI: 10.1111/NPH.14761

[101] Jaganathan D, Ramasamy K, Sellamuthu G, Jayabalan S, Venkataraman G. CRISPR for crop improvement: An update review. Frontiers in Plant Science. 2018;**9**:985. DOI: 10.3389/fpls.2018.00985

[102] Zanella L, Fattorini L, Brunetti P, Roccotiello E, Cornara L, D'Angeli S, et al. Overexpression of AtPCS1 in tobacco increases arsenic and arsenic plus cadmium accumulation and detoxification. Planta. 2016;**243**:605-622. DOI: 10.1007/s00425-015-2428-8

[103] Guo J, Dai X, Xu W, Ma M. Overexpressing GSH1 and AsPCS1 simultaneously increases the tolerance and accumulation of cadmium and arsenic in *Arabidopsis thaliana*. Chemosphere. 2008;**72**:1020-1026. DOI: 10.1016/J.CHEMOSPHERE.2008.04.018

[104] Kumar S, Trivedi PK. Glutathione S-transferases: Role in combating abiotic stresses including arsenic detoxification in plants. Frontiers in Plant Science. 2018;**9**:751. DOI: 10.3389/fpls.2018.00751

[105] Kumar S, Asif MH, Chakrabarty D, Tripathi RD, Dubey RS, Trivedi PK. Expression of a rice lambda class of glutathione S-transferase, OsGSTL2, in *Arabidopsis* provides tolerance to heavy metal and other abiotic stresses. Journal of Hazardous Materials. 2013;**248-249**:228-237. DOI: 10.1016/J.JHAZMAT.2013.01.004

[106] Rajkumar M, Ae N, Prasad MNV, Freitas H. Potential of siderophore-producing bacteria for improving heavy metal phytoextraction. Trends in Biotechnology. 2010;**28**:142-149. DOI: 10.1016/J.TIBTECH.2009.12.002

[107] Mesa V, Navazas A, González-Gil R, González A, Weyens N, Lauga B, et al. Use of endophytic and rhizosphere bacteria to improve phytoremediation of arsenic-contaminated industrial soils by autochthonous *Betula celtiberica*. Applied and Environmental Microbiology. 2017;**83**. DOI: 10.1128/AEM.03411-16

[108] Lampis S, Santi C, Ciurli A, Andreolli M, Vallini G. Promotion of arsenic phytoextraction efficiency in the fern *Pteris vittata* by the inoculation of As-resistant bacteria: A soil bioremediation perspective. Frontiers in Plant Science. 2015;**6**:80. DOI: 10.3389/FPLS.2015.00080

[109] Sessitsch A, Kuffner M, Kidd P, Vangronsveld J, Wenzel WW, Fallmann K, et al. The role of plant-associated bacteria in the mobilization and phytoextraction of trace elements in contaminated soils. Soil Biology and Biochemistry. 2013;**60**:182-194. DOI: 10.1016/j.soilbio.2013.01.012

[110] Shi S, Wang T, Chen Z, Tang Z, Wu Z, Salt DE, et al. OsHAC1;1 and OsHAC1;2 function as arsenate reductases and regulate arsenic accumulation. Plant Physiology. 2016;**172**:1708-1719. DOI: 10.1104/pp.16.01332

[111] Xu J, Shi S, Wang L, Tang Z, Lv T, Zhu X, et al. OsHAC4 is critical for arsenate tolerance and regulates arsenic accumulation in rice. The New Phytologist. 2017;**215**:1090-1101. DOI: 10.1111/nph.14572

[112] Duan G-L, Hu Y, Schneider S, McDermott J, Chen J, Sauer N, et al. Inositol transporters AtINT2 and AtINT4 regulate arsenic accumulation in *Arabidopsis* seeds. Nature Plants. 2016;**2**:15202. DOI: 10.1038/nplants.2015.202

[113] Ali W, Isner JC, Isayenkov SV, Liu W, Zhao FJ, Maathuis FJM. Heterologous expression of the yeast arsenite efflux system ACR3 improves *Arabidopsis thaliana* tolerance to arsenic stress. The New Phytologist. 2012;**194**:716-723. DOI: 10.1111/j.1469-8137.2012.04092.x

[114] Duan G, Kamiya T, Ishikawa S, Arao T, Fujiwara T. Expressing ScACR3 in rice enhanced arsenite efflux and reduced arsenic accumulation in rice grains. Plant & Cell Physiology. 2012;**53**:154-163. DOI: 10.1093/pcp/pcr161

[115] Li R-Y, Ago Y, Liu W-J, Mitani N, Feldmann J, McGrath SP, et al. The rice aquaporin Lsi1 mediates uptake of methylated arsenic species. Plant Physiology. 2009;**150**:2071-2080. DOI: 10.1104/pp.109.140350

[116] Zhao F-J, Ago Y, Mitani N, Li R-Y, Su Y-H, Yamaji N, et al. The role of the rice aquaporin Lsi1 in arsenite efflux from roots. The New Phytologist. 2010;**186**:392-399. DOI: 10.1111/j.1469-8137.2010.03192.x

[117] Ma JF, Yamaji N, Mitani N, Xu X-Y, Su Y-H, McGrath SP, et al. Transporters of arsenite in rice and their role in arsenic accumulation in rice grain. Proceedings of the National Academy of Sciences. 2008;**105**:9931-9935. DOI: 10.1073/pnas.0802361105

[118] Chen Y, Xu W, Shen H, Yan H, Xu W, He Z, et al. Engineering arsenic tolerance and hyperaccumulation in plants for phytoremediation by a PvACR3 transgenic approach. Environmental Science & Technology. 2013;**47**: 9355-9362. DOI: 10.1021/es4012096

[119] Song W-Y, Yamaki T, Yamaji N, Ko D, Jung K-H, Fujii-Kashino M, et al. A rice ABC transporter, OsABCC1, reduces arsenic accumulation in the grain. Proceedings of the National Academy of Sciences. 2014;**111**:15699-15704. DOI: 10.1073/pnas.1414968111

[120] Hayashi S, Kuramata M, Abe T, Takagi H, Ozawa K, Ishikawa S. Phytochelatin synthase OsPCS1 plays a crucial role in reducing arsenic levels in rice grains. The Plant Journal. 2017;**91**:840-848. DOI: 10.1111/tpj.13612

[121] Qin J, Rosen BP, Zhang Y, Wang G, Franke S, Rensing C. Arsenic detoxification and evolution of trimethylarsine gas by a microbial arsenite S-adenosylmethionine methyltransferase. Proceedings of the National Academy of Sciences. 2006;**103**:2075-2080. DOI: 10.1073/pnas.0506836103

[122] Shikha V, Pankaj Kumar V, Veena P, Rudra Deo T, Debasis C. Transgenic *Arabidopsis thaliana* expressing fungal arsenic methyltransferase gene (WaarsM) showed enhanced arsenic tolerance via volatilization. Environmental and Experimental Botany. 2016;**132**:113-120. DOI: 10.1016/J.ENVEXPBOT.2016.08.012

[123] Cao Y, Sun D, Chen J-X, Mei H, Ai H, Xu G, et al. Phosphate transporter PvPht1;2 enhances phosphorus accumulation and plant growth without impacting arsenic uptake in plants. Environmental Science & Technology. 2018;**52**:3975-3981. DOI: 10.1021/acs.est.7b06674

[124] Das N, Bhattacharya S, Bhattacharyya S, Maiti MK. Identification of alternatively spliced transcripts of rice phytochelatin synthase 2 gene OsPCS2 involved in mitigation of cadmium and arsenic stresses. Plant Molecular Biology. 2017;**94**:167-183. DOI: 10.1007/s11103-017-0600-1

[125] Fernandes AP, Holmgren A. Glutaredoxins: Glutathione-dependent redox enzymes with functions far beyond a simple thioredoxin backup system. Antioxidants & Redox Signaling. 2004;**6**:63-74. DOI: 10.1089/152308604771978354

[126] Verma PK, Verma S, Pande V, Mallick S, Deo Tripathi R, Dhankher OP, et al. Overexpression of rice glutaredoxin OsGrx_C7 and OsGrx_C2.1 reduces intracellular arsenic accumulation and increases tolerance in *Arabidopsis thaliana*. Frontiers in Plant Science. 2016;**7**:740. DOI: 10.3389/fpls.2016.00740

[127] Tripathi RD, Tripathi P, Dwivedi S, Kumar A, Mishra A, Chauhan PS, et al. Roles for root iron plaque in sequestration and uptake of heavy metals and metalloids in aquatic and wetland plants. Metallomics. 2014;**6**:1789-1800. DOI: 10.1039/c4mt00111g

[128] Wu C, Huang L, Xue S, Pan W, Zou Q, Hartley W, et al. Effect of arsenic on spatial pattern of radial oxygen loss and iron plaque formation in rice. Transactions of the Nonferrous Metals Society of China. 2017;**27**:413-419. DOI: 10.1016/S1003-6326(17)60047-5

[129] Wang H, Xu Q, Kong YH, Chen Y, Duan JY, Wu WH, et al. *Arabidopsis* WRKY45 transcription factor activates PHOSPHATE TRANSPORTER1;1 expression in response to phosphate starvation. Plant Physiology. 2014;**164**:2020. DOI: 10.1104/PP.113.235077

[130] Castrillo G, Sánchez-Bermejo E, de Lorenzo L, Crevillén P, Fraile-Escanciano A, Tc M, et al. WRKY6 transcription factor restricts arsenate uptake and transposon activation in *Arabidopsis*. Plant Cell. 2013;**25**:2944-2957. DOI: 10.1105/tpc.113.114009

[131] Wang F-Z, Chen M-X, Yu L-J, Xie L-J, Yuan L-B, Qi H, et al. OsARM1, an R2R3 MYB transcription factor, is involved in regulation of the response to arsenic stress in rice. Frontiers in Plant Science. 2017;**8**:1868. DOI: 10.3389/fpls.2017.01868

[132] Huang Y, Chen H, Reinfelder JR, Liang X, Sun C, Liu C, et al. A transcriptomic (RNA-seq) analysis of genes responsive to both cadmium and arsenic stress in rice root. Science of the Total Environment. 2019;**666**:445-460. DOI: 10.1016/j.scitotenv.2019.02.281